"Dr. Buie's *Problem Solving for New Engineers* presents a terrific introduction into the realistic experimental workspace and data analysis for new engineers and scientists. This well-written one-stop overview of experiment planning, execution, and data reduction will be a beneficial stepping-off point to anyone entering into the laboratory for the first time, as well as experienced experimenters reviewing what might go (or went!) wrong."

John Paff
Engineering Technology Manager, Spectra-Mat, Inc.

"Finally, a book that cultivates the rich landscape between human creativity and ingenuity, which motivates the scientist and engineer, and the rigors of applied experimental practice. Looking back over many years of engineering research development and manufacturing activities, I am ever surprised how common problem-solving skills and experimental methodologies are infrequently cultivated alongside the prodigious evolution of technical knowledge and our means to generate data and simulate results. A thoughtful and approachable problem-solving primer has long been needed for new engineers, which combines core experimental principles used in engineering, science, and applied statistics. In academic settings, such subjects are still taught as parts of course work across disparate disciplines. But in contemporary industry, their combination becomes a mandatory core skill set and is key to success in the technical quality and communication of any engineer's creative endeavor.

In Buie's book, we have a contemporary amalgamation of applied experimental principles and methods presented in an approachable and motivating format. Dr. Buie draws from history, case studies, and real examples that breathe life into what might otherwise become a dry subject. Her passion for experimental investigation and its teaching is strongly evident as she traverses a subject matter that might take years of academic and industrial practice for an engineer to integrate and master."

Len Mahoney, PhD
Unit Process Engineer, Avago Technologies

"*Problem Solving for New Engineers* offers a way to shape learning gained in school and bridge the gap to becoming a savvy, strategic problem solver, reducing the "groping-in-the-dark" phase of mastering a discipline. This book enables the wisdom of mastery by providing key understandings and methods that are at the heart of an experimental discovery mindset. Approaches to moving fascination and wonder into realized outcomes are based in a context of inquiry, exploration, and discovery that refine disciplined problem-solving by happily traveling the unknown—one experiment at a time."

Diana Hagerty

Project Manager at General Atomics Aeronautical Systems

"Melisa Buie is not only creative in her approach but also utterly aware of the challenges we face as engineers and scientists in practice. As I was going through the pages, I realized that the book mirrors my own experience. I wish something like this has been available when I was starting out."

Saman Choubak, PhD

Senior Research and Development Engineer at PepsiCo

"*Problem Solving for New Engineers*, written by Dr. Melisa Buie, serves the fresh new engineers with plenty of methods required for successful experimentation and process development in modern companies, with focus on, but not limited to, nature sciences.

The problem I observe so frequently with new engineers coming from the university—from how to apply the knowledge about how experiments were performed by others to an efficient setup of our own experiments—is discussed at different levels, and guidance is provided every step of the way, from a collection of the requirements to evaluation and qualification of the new process.

Personally, I most appreciate the balance between the overview of methods in a thorough explanation, rather free of equations, which will not let you skip the rest of any chapter, and a fair comparison of the one-factor-at-a-time experimentation that all of us learned at university and statistical design of the experiment.

The text invites you to experiment on your own and irradiates the pleasure of investigation and development itself. The author's knowledge of science history converts the scientific topic to an easy-to-read lecture, which you will also enjoy as a bedtime story."

Pavel Nesladek, PhD

Member of Technical Staff, Advanced Technology Mask Center

"I wish I had this book when I was a college student! I might have decided to become an engineer or a scientist. Melisa Buie brings her background in industry and academia together in a balanced and effective way. This is not your usual dry textbook. I laughed out loud in places! It's a must-have "how-to" reference book that is focused on important engineering and scientific concepts, communicating experiments and research effectively, and being a successful engineer or scientist in academia or private industry. The material is presented in an exciting, real, and sometimes humorous way by using stories, sharing life experiences, and revisiting discoveries of the great scientists throughout history. As someone who has been responsible for recruiting new engineers fresh out of college, reading this book should be a prerequisite for being hired."

Noël Kreidler
Owner, Kreidler Solutions, Talent Acquisition and Human Resources

"In today's fast-paced technology industry, being able to efficiently attack issues and clearly share learning is critical. As Dr. Buie points out, many engineers entering the workforce have a strong science background, but their real-life problem-solving skills are not as developed. This book is a wonderful overview of problem-solving strategies, experiment design, and data analysis needed to succeed in a world driven by constant discovery. I especially appreciate the sections on graphing, as poorly communicated learning within cross-functional teams can lead to wasted time and effort down the road. This text should be required reading for all newly hired engineers and a welcome reference for those of us who have worked in this industry for many years."

Jeremiah Pender, PhD
Senior Engineering Development Manager, Applied Materials, Inc.

Problem Solving
for New Engineers

What Every Engineering Manager
Wants You to Know

Problem Solving
for New Engineers

What Every Engineering Manager
Wants You to Know

Melisa Buie

CRC Press
Taylor & Francis Group
Boca Raton London New York

CRC Press is an imprint of the
Taylor & Francis Group, an **informa** business

A PRODUCTIVITY PRESS BOOK

CRC Press
Taylor & Francis Group
6000 Broken Sound Parkway NW, Suite 300
Boca Raton, FL 33487-2742

© 2018 by Melisa Buie
CRC Press is an imprint of Taylor & Francis Group, an Informa business

International Standard Book Number-13: 978-1-138-19778-7 (Hardback)

Library of Congress Cataloging-in-Publication Data

Names: Buie, Melisa, author.
Title: Problem solving for new engineers : what every engineering manager wants you to know / Melisa Buie.
Description: Boca Raton : Taylor & Francis, CRC Press, 2017. | Includes bibliographical references.
Identifiers: LCCN 2017003837 | ISBN 9781138197787 (hardback : alk. paper)
Subjects: LCSH: Engineering--Vocational guidance.
Classification: LCC TA157 .B835 2017 | DDC 658.4/0302462--dc23
LC record available at https://lccn.loc.gov/2017003837

Visit the Taylor & Francis Web site at
http://www.taylorandfrancis.com

and the CRC Press Web site at
http://www.crcpress.com

"If I have seen further it is by standing on the

shoulders of giants." – Isaac Newton, 1676

For my son, Benjamin Clay Alexander-Buie,

and my parents, John and Mary Buie.

Giants may be a myth to some, but in my eyes and my heart, these three

people are giants. I have learned so much from them, more than I write.

Contents

Foreword

As engineering students transition into engineers in industry, many learn that their new skills are inadequate to answer a variety of the design decisions they face. The world is more complicated and system behavior is more subtle than can be worked out with basic engineering calculations. Two of the greatest skills needed in industry are how to make trial and error more efficient and effective and how to cope with variation. Making trial and error more efficient and effective is the domain of experimental design; coping with variation is the domain of statistical methods. By combining the two, a model of system behavior is built. Yet most engineering students have not had a course in experimental design and, typically, just a very introductory course in statistical methods, one that does not cover complex model fitting.

Trial and error (or hypothesize and test) is the scientific method. For a complex process that depends on a number of factors, the only way to understand and model the process behavior is with a multifactor experiment. The field of experimental design demonstrates how to learn system behavior in the most efficient way: a way that holds outside factors constant, that helps you understand interactions between factors, and that allows you to learn many things at once rather than just one factor at a time.

With statistical methods, process variation becomes clear. The data coming from monitoring a process need to be studied statistically to adequately judge when the system behavior is changing, rather than simply exhibiting natural variation. We live in an age of omnipresent data; statistical methods provide the tools to understand what the data are revealing.

But here is the disconnect. Despite the overwhelming value of experimental design and statistical methods, they are not being sufficiently taught in most engineering curricula.

BREAKTHROUGHS I WANT YOU TO KNOW ABOUT

There are a few gotchas and game changers I want you to become aware of as you learn to model using modern software. Here are my top three. Please remember them; I include them here because they are relatively new and thus not covered in most texts.

1. *Optimal design of experiments.* Using computer-based optimization is the modern way to construct an experimental design. It will give you the best design for your situation: for your run budget, for your factor restrictions, for what you need to be able to estimate, and for any combination of factors, including two-level, many-level, mixtures, and blocking. The new way is easy and optimal, and it is no longer worth learning the old ways in traditional textbooks. Suppose you have three mixture factors, several categorical factors in addition to a couple of continuous factors, with blocks of size five and a run budget of 35. Textbook designs won't be much help, but optimal design will handle it easily.

2. *Recognizing a split plot experiment.* Very often in experiments, some of the factors are hard to change, while others are easy to change. By grouping the runs that have the same settings for the "hard-to-change" factors, the experiment becomes easier to run. This grouping introduces a different statistical structure—a split plot—to the experiment. The optimal design of the experiment needs modern software to deal with the split plot structure, and the fair and efficient estimation of the model also needs modern software (REML with Kenward-Roger bias adjustments, if you want to know). Failure to recognize the split plot structure of an experiment can lead to bad conclusions.

3. *Definitive screening designs.* When designing a screening experiment to identify the important factors, interactions may intrude. Older experimental designs (like lower-resolution fractional factorials) suffer when there are strong interactions that are not part of the model, resulting in biases in the main effects estimates. The definitive screening design was the breakthrough that fixed this problem by making the main effects orthogonal to the interactions (second-order effects). In addition, the definitive screening design has some

ability to reveal curvature and two-factor interactions in a model. And you get all these benefits with just a small sacrifice in main-effect efficiency.

IN THE REAL WORLD

It is a very competitive, challenging world economy out there. In big companies, often their flagship products were invented long ago, meaning that the original patents have expired and the competitors have had time to get very good. It becomes a race to get the last ounce of performance, to eke out the next opportunities to save on cost, to adapt to the next iteration of design, to pursue the next opportunities to add value. And when the volume is high, the incremental improvements can be worth millions, even billions, even deciding whether or not the company survives its competition. Developing those improvements usually means conducting well-designed experiments and understanding the process behavior in detail.

The newer innovations also exist in a very challenging environment. Each innovation comes with different problems to solve, many of which depend upon a designed experiment to show the best path.

FILLING THE GAP

Sadly, much of the scientific and engineering curricula in many universities around the world fail to adequately prepare students to think about the data they need and the statistical methods required, such as the ones described earlier, for them to be more effective problem solvers. As a result, this gap is most often filled through internal training within the organization or via professional education. Thankfully, there are a few educators who see the need to prepare new scientists and engineers with proven problem-solving skills. Melisa Buie is one of them, which she has demonstrated as an instructor at San Jose State University, as a mentor to fellow professionals within industry, and now as the author of this wisdom-filled book.

Melisa is well suited to impart valuable advice to new engineers and scientists who seek to be better problem solvers. Her experience includes many scientific and engineering accomplishments in her years as a researcher and in her applications in industry. She credibly shares why and how we should use the power of information-rich data from smart experimentation. Though targeted to engineers and scientists, this book benefits anyone who wants to be a better problem solver.

Effective experiments are ultimately about value creation: increasing productivity, minimizing waste, and/or improving quality. These are important goals in any organization. The methods for determining *how* to achieve such goals are readily accessible, having evolved from slow and limited one-factor-at-a-time, trial-and-error experiments (as Melisa says, not a bad way to get our feet wet) to powerful, elegant methods that reveal valuable insights while saving money and time.

Since we are innately curious, Melisa recommends that we satisfy our curiosity with *intentional* experimentation. When you intentionally experiment, you filter out the noise in data, making your efforts more efficient and effective. Intentional experimentation is the journey to fruitful discovery.

For those who are new to the idea of intentional experimentation, this may seem uncomfortable at first. Through engaging analogies, stories, and visualizations, Melisa guides readers so they can see what is required for effective problem solving, empowering them to make their own discoveries and to experiment with confidence. She likens experimental discovery to Scrabble: to do well in a game of strategy like Scrabble, you must know how the game is played while simultaneously strategizing and adapting with the tiles you have.

Melisa's own curiosity is evident throughout this book. As she guides the reader along the path of fruitful discovery—sharing many interesting and time-honored references—her curiosity becomes contagious, making the journey with her enlightening, entertaining, and inspiring. Anyone aspiring to be a more efficient and effective problem solver will find this book to be a useful guide.

To paraphrase Douglas Montgomery, Regents' Professor of Industrial Engineering and Statistics at Arizona State University, engineers and scientists are already smart; we just want to make them smart experimenters so they can learn faster. I applaud Melisa's efforts to help more scientists and engineers be better data detectives so they can discover for themselves the value of applying statistics and experimental investigation.

In the following pages, let a skilled master show you how to apply key statistical concepts so that you can experience firsthand the rewards of discovery and creative problem solving. Enjoy!

John Sall
Co-founder of SAS and Chief Architect of JMP

Author

Melisa Buie, PhD, makes lasers and solves problems. In her role as director of operations, she works on both engineering and business problems. She joined Coherent and began lecturing at San Jose State University in 2007. She has also worked as a research scientist for Science Applications International Corporation, working at the Naval Research Laboratory in Washington, DC, where she made theoretical lasers. Melisa was a member of the technical staff and engineering manager at Applied Materials, Inc., prior to joining Coherent.

Melisa has coauthored more than 40 publications and holds five patents. Melisa's degrees include a PhD in nuclear engineering/plasma physics from the University of Michigan and an MS in physics from Auburn University. She has a Six Sigma Black Belt from the American Society for Quality. In 2017, she completed a certificate in innovation leadership from Stanford University Graduate School of Business. She lives in Palo Alto, California.

1

The Great Universal Cook-Off

The object of science may be said to be to construct theories about the behavior of whatever it is that the science studies. Observation and experience, inspiration and serendipity, genius and just good guesses—by their presence and absence, in pinches and dashes—all can provide the recipes, in which the scientist provides the inexpressible human flavor. This aspect of science, the concoction of theories, has no universal method. But once a theory has evolved, perhaps from a half-baked idea to a precise and unambiguous statement of the scientist's entry in the great universal cook-off, the scientific method may be used to judge the success or failure of a given theory or the relative merits of competing theories.

Henry Petroski

All science begins with problems, and we all use essentially the same method to solve problems. We try things out, we experiment. We put things to the test. Our schools and universities give us the basic knowledge in the fields in science and engineering. We read about others experiments. We learn the results of their tests and trials. But when do we have the opportunity to discover? Our lab classes are intended to open our eyes and have us see what those who've come before us saw. Yet, they often fall short. Our lab classes have us follow detailed instructions with a well-characterized, very limited problem statement. Unfortunately, this is not how problems and experiments occur in real life. The aim of this book is to provide a strategy and the tools needed for new engineers and scientists to become apprentice experimenters armed only with a problem to solve and some knowledge of their subject matter.

1.1 DISCOVER FOR YOURSELF

Our high school and undergraduate (and possibly even some graduate) lab classes have explicit step-by-step instructions on how to run the experiment for that lab. These instructions tell us what to measure, how to measure, when to measure; they tell us how much gas or chemicals to add and when to turn the power on. We are told when to pay attention and when to make observations. We are even told what to observe. In some cases, the detailed instructions provide blank spaces for us to write our observations. For all practical purposes, our undergraduate science and engineering labs are cooking classes. We follow recipes.

Everyone learning how to cook must start with recipes. In the 2009 movie *Julie and Julia*, Julie religiously follows the recipes of Julia Child on her way to becoming an expert (Ephron 2009). In an effort to create dishes similar to what Julia Child would have prepared in France in the 1950s, a novice cook would be wise to follow instructions from her culinary classic, *Mastering the Art of French Cooking* (Child 1961). Recipes handed down to us from masters or relatives are blueprints we can use to help us begin. When our parents or grandparents pass down their secret recipe for our favorite dish from childhood, we make every attempt to duplicate it. We practice by imitation their precise movements. We go deep into our memory banks and recall how Grandma measured the flour or how Dad tilted the dish to stir. If it doesn't turn out exactly correct, and we call home describing the failure, we'll get off the phone with another set of instructions on how to properly prepare the dish. Eventually, if we keep working at it, we can make Mom's pound cake, Dad's banana pancakes, or Granddad's chili just like they did.

We use recipes to validate what we already know. We use recipes to make the same thing in the same way each time. We use recipes to build our confidence. Before we can cook like Paula Deen, Gordon Ramsay, Julia Child, or grandmother, we mimic them by following their recipes. It is by mimicking the great artists initially that we can begin to learn. Mimicking prepares our mind for mastery.

What happens when we deviate? What happens when we use whole wheat flour rather than white flour in a pound cake? What if we add chocolate chips to those banana pancakes? When we decide to alter recipes rather than just duplicate them, we're forced to ask questions. Through our own successful and sometimes unsuccessful permutations, we discover the limits of our own abilities and invent creative solutions for each new problem we encounter.

What does this have to do with science and engineering? Everything! Science classrooms are designed to teach us the universal discoveries and the work of those who came before us. Galileo, Johannes Kepler, Isaac Newton, Marie Curie, and Wilbur and Orville Wright are just a few of the many whose work has changed our understanding of the world. We study the world of knowledge they have discovered. How did they begin to discover for themselves? How do we begin to step beyond the world they mapped out for us?

Although many imagine experimentation to involve lab scientists with white coats and expensive equipment, experimentation is as simple as asking the question "What if I made one small change?" Experimenting in its most basic form means asking, then acting upon, questions. We wonder about something, try it out, and evaluate the results (Berger 2014).

My earliest experiments came from my family, farmers in the southeastern United States. My first lab was the family farm. Although my grandfather had only a fourth grade education, he religiously watched the *Noon Farm Report*. One of my earliest memories is watching Gene Reagan provide updates on the latest experiments in agricultural science. The *Noon Farm Report*, hosted by Gene Regan, was broadcast each weekday at noon on the local station, WTVY (WTVY 2016). The farmers would watch the show, and I would watch with them, until soon enough I was inventing experiments left and right. I recall asking my father to help me figure out something useful to do with the Japanese vine, kudzu. The vine was introduced to control erosion in the south, but the invasive species grew over everything in its path, from houses to trees to mountains. I thought it would be beneficial to invent some new use for this vine that grows like wildfire. These experiments were all a bust; this invasive, green monster was not to be converted to anything useful by me. I've since learned that others have tried their hand at it with more success. We can now purchase jelly or put the leaves in our salad.

When we were young, most of us had inquisitive, experimental minds. Now we just don't think of life in those terms. Observing some anomaly we'd like to test or understand more deeply, then asking a question about it, is the beginning of experimentation. "If you randomly try things in life, it can lead to haphazard results; but if you take time to consider why they might be worth trying, and what might be the best way to test them out, and then assess whether the trial was a success …" then we are experimenting (Berger 2014).

Engineers and scientists do not want to spend their careers following step-by-step instructions. We are not hired to spend our careers following recipes. In industry, we hire engineers and scientists to solve problems by performing experiments, analyzing data, then developing and implementing solutions. Within our current education paradigm, the expectation is that we begin by following the instructions in our labs. Professor Truran of the University of Minnesota wrote that experimentation in science is skewed "toward the acquisition, demonstration and utilization of theoretical knowledge. Experiments rarely represent an opportunity for testing of hypotheses, but typically take the form of demonstrations.... There will be limited insight into the inventiveness, persistence and technical creativity that are necessary to carry out an important experiment" (Truran 2013). At the very best, our science labs in college provide step-by-step instruction (recipes) for how to validate some known result. Assuming we follow the instructions precisely, we should validate for ourselves that expected experimental result. High-tech manufacturing is a perfect example of an applied science/engineering application where there are recipes (cookbooks), and for the same reasons. Our companies want to build the same computer chip, laser, television, or drug, over and over. These recipes are necessary to provide control over the process. In an ideal world, in order to manufacture a product, cook a Hollandaise sauce or measure centripetal force, the recipe would be so simple that anyone could follow it, even if that person might not understand every intricacy of the process or equipment.

Ultimately, the scientist or engineer is responsible for developing these recipes and transferring them to others. How do we go from following the recipe for *reine de saba* (chocolate and almond cake) to becoming the next Julia Child? At what point do we transition from validating other people's work to creating our own? How do we go from following the step-by-step instructions provided in our labs in undergraduate courses to writing these instructions for others? How do we move to creating and designing our own experiments? How do we make the work of science and engineering our own? My goal is that the succeeding chapters ease the transition into independent investigation. First, let's consider motivation.

1.2 CREATING A CONTEXT FOR DISCOVERY

Inspiration is needed in geometry, just as much as poetry.

Pushkin

Many of us desire to become scientists and engineers because of a love, desire, or passion to discover new things or solve problems. For some of us, there is a particular problem that motivates us, that provides the context in which we desire to grow. One of my best friends in college studied plasma physics because she wanted to help give the world fusion energy, the ultimate safe, clean energy source. For another friend, her life changed when she found astronomy. She is now designing and building some of the most sophisticated detection equipment for astronomical exploration. John Steininger, founder of tech start-up Sonopro, wanted to "light up Africa." Through John's work, microfinanced solar powered lamps provide lights for students in Uganda. Thane Kreiner, executive director and professor of Science and Technology for Social Benefit at Santa Clara University, uses his neuroscience and business background to solve world problems like bringing fresh drinking water to remote locations.

My first visit to Ann Arbor, Michigan, was Earth Day 1990. There was a campus-wide festival. A positive upbeat atmosphere abounded. My initial meeting was with Professor Ronald Fleming, who ran the Ford Nuclear Reactor on the university's North Campus. As Professor Fleming walked me through the reactor, he told me about his visits to India, where he saw people starving a short distance away from food. The problem seemed to be how to transport the food from its source to these people while keeping it fresh. Professor Fleming was passionate about developing a technique using irradiation as a solution to preserving the food until it could reach the people in need. Irene Joliot-Curie, Nobel Prize winning chemist and daughter of Marie and Pierre Curie, felt strongly that "nuclear energy has only one objective, the improvement of the economy of our daily lives" (Goldsmith 2005).

I have always loved solving problems and talking about science with others who are passionate about the world. Discovering our context for experimentation, whatever gives us that spark or drives us to discover, is an important part of the process. In the words of Claude Levi-Strauss (1983), "The scientific mind does not so much provide the right answers as ask the right questions." Author Ian Leslie writes, "Questions weaponize our curiosity, turning it into a tool" (Leslie 2014). When we begin to ask the right questions out of intellectual curiosity or out of a passion to solve a local or global problem, then we have taken the first step toward a scientific mind.

Discovery is defined as "the action or an act of finding or becoming aware of for the first time, especially the first bringing to light of a scientific phenomenon" (Brown 1993). The word *experiment* has multiple definitions: (a) the action of trying something or putting it to the test; a test, a trial; (b) an action or procedure undertaken to make a discovery, test a hypothesis, or demonstrate a known fact; (c) a procedure or course of action tentatively adopted without being sure that it will achieve the purpose; (d) ascertain or establish by trial; and (e) experience (Brown 1993).

Experimentation, whether we are following a recipe, modifying a recipe, or exploring a new idea altogether, is a means for scientists to discover for themselves the mysteries of the world. Experimentation is a very intentional imposition of conditions on a sample or population of interest where a response or responses are observed (Easton and McColl 2016). Experimentation and problem solving have us study what happens when conditions changes. "In experimentation, unlike pure observation, the person doing the experiment deliberately changes different parameters, often to clarify cause and effect" (Nobel 2016). Physics Professor Richard Feynman said that he could never really understand someone else's work until he had done it himself (Feynman 1965). The aim of this book is to provide a strategy and the tools needed for beginning experimenters who follow recipes with detailed instructions to become apprentice experimenters armed only with a problem to solve and some knowledge of their subject matter.

As investigators, we want to create experiments to answer our questions and address our curiosities. Marie Curie wrote, "I am among those who think that science has great beauty. A scientist in his laboratory is not only a technician, he is a child placed before natural phenomena, which impress him like a fairy tale" (Labouisse 1937). Professor Richard Feynman wrote, "The principle of science, the definition, almost, is the following: The test of all knowledge is experiment. Experiment is the sole judge of scientific truth" (Feynman 1977).

1.3 REQUIREMENTS FOR EXPERIMENTAL DISCOVERY

The requirements for experimental discovery are very similar to the requirements for playing any game of strategy. The game of Scrabble has each player pull seven lettered tiles with different point values, create words from the tiles in hand, and connect them to the tiles on the board. Points are given for the letter values, word value, and strategic placement on the board. The objective of the game is to have the most points when the tiles run out. Consider the game Twenty Questions, or as some of you may know it, Animal, Mineral, Vegetable. The objective of the game is to identify the object that a person is thinking of by asking up to 20 yes/no questions. Take a break from reading and play a couple of rounds with friends, classmates, or colleagues.

Think about the game. What do you observe from playing these games? Did you notice that there is not a unique route to the problem solution? Two, three or four equally competent players presented with the same problem might typically begin from different starting points, proceed by different routes, and yet could reach the same answer. What is sought is not uniformity (doing everything exactly the same) but convergence (reaching the same answer).

Notice that both games, Scrabble and Twenty Questions, follow an iterative pattern. Each hand of tiles or new question is like a newly designed experiment. Each new hand of tiles can be arranged and rearranged in conjunction with the tiles existing on the board. At each stage of Twenty Questions, the hypothesis is progressively refined and leads to another question (experimental design) that elicits data that leads to a modification of the hypothesis.

What does it take to play these games well? There are two essential elements: knowledge of subject matter and knowledge of strategy. The strategy for Twenty Questions is well known. At each stage, a question should be asked that divides the objects not previously eliminated into equally probable halves. Did you use this strategy at least once? Did you think of it as a strategy?

What lesson can we learn from these simple games? In Twenty Questions, it would be difficult to arrive at the answer "*Nina, Pinta,* and *Santa Maria*" if we do not know much about early explorers or Christopher Columbus or to get "Tina Fey" if we do not know anything about comedians or *Saturday Night Live.* Likewise, it would be difficult in Scrabble to create the seven-letter word "eremite" if we didn't even know it was a word. (This happened to me.) Did you notice that without knowledge of the appropriate strategy, it is still possible to play the game, just maybe not as well? However, if you do not have knowledge of the subject matter, it is almost impossible to play. The best game players are those who have knowledge of both strategy and subject matter. Professor Weisberg wrote, "Creative thinking begins with what is already known" and inventive problem solving is "based on a near analogy to the solution the inventor is dealing with." When new problems arise, creative thinking then "goes beyond the already known. ... Even the most talented individuals, in order to produce influential work, must acquire expertise in a domain.

This acquisition serves the creative process as the basis for continuity in thinking, raising the question if, in order to begin to do innovative work in a domain, one must know what came before" (Weisberg 1993). Knowledge of subject matter is the key to getting started. Developing a knowledge of strategy is essential to efficient, consistent experimentation and problem solving.

We need a certain amount of information before we can begin experimentation. First, we want to know what others know. We want to know the terms, jargon, tools, history, and anecdotes in our field of study. When new content is introduced, it should fill in the gaps and add to our existing knowledge and mental models. We examine the soundness of what we already know. We must constantly ask ourselves, "How confident am I in this information?" We need to understand how that knowledge constrains, shapes, and distorts us. There are times when "what we think we know" keeps us from asking the questions necessary. However, it is critical we understand what others have done and what the experts know about the area we are experimenting in. We can use this learning to critically reflect on the physical world. This knowledge of terms, jargon, tools, history, and assumptions in our field of study will provide us with openings for action (questions and curiosities) that were previously unavailable to us within the constraints of our existing model of the world. We move back and forth between gathering information about our subject, to the unknown, all the while increasing what we know.

Knowledge of strategy in these gaming examples parallels the strategic knowledge that we need in scientific investigations. It is important to understand which strategic tools to use in the experiment. Dr. Khorasani divides this strategy into data analysis and statistical knowledge and thinking. By combining these pieces of the strategy with our knowledge of subject matter, we can begin to explain the results of our investigations. A scientist or engineer can conduct an investigation without statistics, but it is impossible for the experiment to be performed objectively and efficiently without an understanding of statistics (Bode et al. 1986, Boring 1919, Box et al. 1978, Khorasani 2016). A good scientist or engineer becomes much better with the knowledge of strategy. This is particularly true in fields with large data sets such as medicine where conclusions have public health implications. As medical doctor Vladica Velickovic wrote, "Involvement of biostatisticians and mathematicians in a research team is no longer an advantage but a necessity" (Velickovic 2015).

1.4 REQUISITE WARNING LABEL

Most scientists are not statisticians. At best, they've had one or two classes in college studying statistics. As a result, inaccuracies due to statistical errors are pervasive. According to an article published in *The Economist* in 2013, an official at the National Institutes of Health estimated that 75% of the published biomedical findings are not reproducible (*Economist* 2013). Glen Begley from Amgen and Lee Ellis from University of Texas M. D. Anderson Cancer Center attempted to replicate the results in 53 classic "landmark" oncology studies. They were able to confirm only six (11%) of these results (Begley and Ellis 2012).

As with a new drug that treats the symptoms of the flu, experimental discovery with a statistical toolbox comes with a few warnings on its label. Neither statistics nor statistical software packages should be used without proper education on what the tools are and how/when to use them appropriately. Statistics are tools for us. The more complex the problem, the more we need to lean on these tools. As with a physical tool, it is important that we understand the subtleties of the statistical tools we use.

The potential benefits to statistical analysis are amazing and powerful. There are three primary sources of difficulty that can be avoided by using statistical methods: (1) quantifying and categorizing variation, (2) demystifying causation with correlation, and (3) unraveling complex interactions (Box et al. 1978). These three areas of inquiry confound many well-known and experienced scientists.

1.4.1 Understanding Variation

The first source of difficulty is understanding variation. Variation is a double-edged sword. Variation happens no matter what we do. We'll spend three chapters with the quantification of variation—variation that adds uncertainty to our data. In our experiments, we only want variation due to the changes we are making to our experimental variables, i.e., the variation we can control. We can't always control variation; therefore, it is critical that we understand the potential sources of variation and plan our experiments with variation in mind. In this book, we will examine data and learn how to minimize and quantify uncertainty due to various types of variation— random fluctuations, mistakes, and systematic bias. Once we've accounted for random and systematic variation (bias) and minimized the potential for mistakes, we can move to intentionally varying process parameters and studying the resultant changes in the response parameters.

1.4.2 Demystifying Causation and Correlation

Correlations everywhere, now I must stop and think.

Steven Novella

A second source of difficulty is confusing correlation with causation. Statisticians and really good scientists are a very cautious and noncommittal bunch. Rather than discuss certainty, we qualify all our statements with a degree of probability. Even with large data sets and replicated experiments, we still speak of causal relationships probabilistically (Randall 2011). We can never prove that two variables are related. The strongest commitment statement that we will hear experimenters make is "there is a strong probability that two quantities are related."

Figures 1.1 and 1.2 show a direct correlation between life expectancy at birth in the United States and the number of patents granted. Could this be the secret to a long life? Of course, this is ridiculous. There are many examples like this one. For example, the number of civil engineering doctorates awarded is directly correlated to the per capita consumption of mozzarella cheese. There is a strong correlation between computer science doctorates and comic book sales, mechanical engineering doctorates, and World of Warcraft subscribers (Vigen 2015). Of course, we know that the more cheese consumed doesn't result in universities issuing more civil engineering doctorates or comic book readership have more computer science doctorates issued. These may seem like silly examples; however, these types of logical fallacies are made each day. Have you ever said, "It never fails, if I wash my car, it will rain" or "As soon as I get in the bath, the phone rings" (Falin 2013)? Remember that we are human first then scientists—"thinking anecdotally comes naturally, whereas thinking scientifically does not" (Shermer 2008).

There are many documented and even more undocumented cases of scientists collapsing correlation and causation. As scientists, we know this and yet we continue to conflate correlation and causation in leading peer-reviewed professional journals. There are multiple fairly recent articles that present a list of well-known, respected scientists who had made a mistake similar to the previous example—confusing correlation with causation (Velickovic 2015, Wainer 2007). Did you hear about the 2012

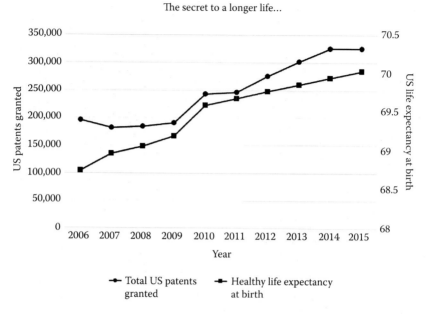

FIGURE 1.1

Graphic that illustrates the correlation between the number of patents granted in the United States and life expectancy in the United States. (From United States Trademark and Patent Office, U.S. Patent Activity Calendar Years 1790 to the Present: Table of Annual U.S. Patent Activity Since 1790, https://www.uspto.gov/web/offices/ac/ido/oeip/taf/h_counts .htm, 2016; World Happiness Report: Overview, http://worldhappiness.report/overview/.)

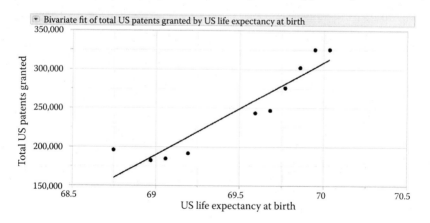

FIGURE 1.2

Bivariate fit of the the number of patents granted in the United States and life expectancy in the United States. (From World Happiness Report: Overview, http://worldhappiness.report /overview/; United States Trademark and Patent Office, U.S. Patent Activity Calendar Years 1790 to the Present: Table of Annual U.S. Patent Activity since 1790, https://www.uspto.gov /web/offices/ac/ido/oeip/taf/h_counts.htm, 2016.)

New England Journal of Medicine study linking chocolate consumption with enhanced cognitive function? The paper was based on the correlation between the countries of Nobel Prize laureates and per capita chocolate consumption in that country (Velickovic 2015).

Proclamations of causation should be used with caution. This logical fallacy is so common and prevalent that there is an official name for it: "Cum hoc ergo propter hoc," which translated means "with this, therefore because of this." We'll look at using regression analysis and designed experimentation to model experimental findings in Chapters 8 and 9. However, with the clarification and quantification of the source of variation in our experiments and a strong knowledge of subject matter, it is possible to express our results with the appropriate level of confidence. Winner of the Nobel Memorial Prize in Economic Sciences, Professor Daniel Kahneman, wrote, "Regression effects are a common source of trouble in research, and experienced scientists develop a healthy fear of the trap of unwanted causal inference" (Kahneman 2011).

1.4.3 Unraveling Complex Interactions

Finally, the third difficulty facing inquiring minds is the complexity of the effects being studied. Let's consider a study of the effects of coffee and chocolate on the race times of relay runners. Suppose it was found that if no coffee were consumed, 1 g of chocolate decreased the race time by an average of 0.45 seconds in the 200-m relay, and if no chocolate was consumed, one cup of coffee reduced the race time by an average of 0.20 seconds. Our job as experimenters would be so much easier if the effects of several grams of chocolate and several cups of coffee and their combined effects were linear and additive. For example, if they were linear, then 2 g of chocolate would reduce the race time by 0.9 seconds, and three cups of coffee would reduce it by 0.60 seconds. If it were linear and additive, then 10 g of chocolate and 15 cups of coffee would reduce the

relay leg time by $\left[10g\left(0.45\dfrac{\text{seconds}}{g} \right) + 15g\left(0.2\dfrac{\text{seconds}}{\text{cup}} \right) \right] = 7.5$ seconds. It

is much more likely that the effects of coffee are not linear; therefore, the effect of one cup of coffee will depend on the amount of chocolate and coffee previously consumed. Actually, it's much more likely that the runner who consumed all this chocolate and coffee would be out of the race altogether. In the final chapters, we will learn how to build models that capture the complex interactions of experimental variables.

1.5 BOOK ORGANIZATION

This book is organized into a series of lessons or essential core concepts that fit together to solve a big-picture problem. Problem solving and experimentation are key elements in the development of new products, new technologies, and new ideas. Our foundation is a solid grasp of engineering and physics principles (knowledge of subject matter). Adding strategic experimental design thinking to this foundation, we can build a solid, repeatable experiment with full awareness of the limitations and strengths of our experimental findings. With the big picture in our sights, each chapter explores critical concepts related to *variation* that build upon one another. The lessons in this book are organized around variation, which is introduced in Chapter 4.

Chapter 2—Eureka! And Other Myths about Discovery

This is an important chapter to begin the book. We have developed some preconceived notions about science and discovery. These myths about scientific discovery, while we may not even be aware of them, can hold us back from the sheer joy of discovery.

Chapter 3—Experimenting with Storytelling

This chapter focuses on the importance of the most common types and forms of communication in science and engineering. Our work is presented in reports or memos, journal articles, posters, and in oral presentations. We communicate our work in written, tabular, and visual displays. In this chapter, we examine ideas related to embracing the storytelling of science in writing and speaking, the purpose of tables, and how to get the most out of visual presentations of data—charts, photographs, micrographs, schematics, etc.

Chapter 4—Introducing Variation

In this chapter, we review the properties and characteristics of data and uncertainty that we've learned in school. Uncertainty leads into our discussion of variation. This foundational chapter sets the stage for the next three chapters, which cover unintentional variation, systematic variation, and random variation.

Chapter 5—Oops! Unintentional Variation

Experimentation is full of opportunities for mistakes which can result in unintentional variation in our results. In this chapter, we will look at some mistake proofing tools like checklists, Standard Operating Procedures, and Input–Process–Output diagrams to minimize unintentional variation.

Chapter 6—What, There Is No Truth?

In this chapter, we look at the contribution of variation from the measurement system using measurement system analysis. Variation attributable to a measurement system is quantifiable. We'll look at characterization and quantification of that variation.

Chapter 7—It's Random, and That's Normal

Variation happens. It's normal and it's Normal! The *Normal Distribution* is introduced in this chapter but not in the normal way. This powerful and amazing tool is used universally in all disciplines of science and engineering. We can learn so much from this very simple tool.

Chapter 8—Experimenting 101

The alternate title for this chapter might be "Intentional Variation: One-Factor-at-a-Time." Now that we know how to characterize and account for unintentional variation and quantify random variation and systematic bias in our measurements, we can intentionally introduce variation into an experiment so that we can characterize it. In this chapter, we examine the familiar and comfortable practice of one-factor-at-a-time experimentation. We'll build on solid algebraic techniques to build mathematical models of our physical system.

Chapter 9—Experimenting 201

The alternate title for this chapter might be "Intentional Variation: Designed Experimentation." It is possible to perform experiments where we change multiple input parameters at the same time. In this chapter, we extend the concepts from Chapter 8 into multiple variable experimentation.

Chapter 10—Strategic Design: Bringing It All Together

In the chapter, we pull it all together. We will have spent five of the prior nine chapters outlining different types of variation and how to control it.

We'll walk through how to encapsulate all these pieces in a coherent strategy and solid experimental plan.

Chapter 11—Where to Next?

This chapter contains concluding remarks; but if you looked at the outline, you know that there are actually 12 chapters.

Chapter 12—One More Thing...

I couldn't close this introductory text without providing additional references on topics that are covered here. There are many more references than those listed in Chapter 12, but it's a place to start.

1.6 KEY TAKEAWAYS

Today, we can stand on the shoulders of many giants that have come before us. Duplicating and mimicking the results of others are great ways to begin experiencing experimentation. Using recipes created by the scientists and engineers who preceded us allows us to learn about the subject. We will learn by doing these experiments for ourselves and perhaps repeating some of them. "Graduate students could, in addition to learning the guidelines, train by replicating published studies" (Fanelli 2013). Experimentation builds experiential muscles that no amount of reading what others have done can give us and that no one can take from us.

As scientists or engineers with a solid foundation in physics and engineering principles and a few statistical tools, we should be able to begin experimental exploration and discovery for ourselves. My goal in this book is that we come away knowing how to begin to discover for ourselves through experimental investigation. Reading and working through this book, we will become fully equipped with the tools, skills, and fearlessness required to discover for ourselves those things that may be known to others or may not be known at all. With a lot of patience, knowledge, strategy, and a bit of luck, we may discover something previously unknown to anyone.

P.S. Take some time to explore the ideas in this chapter. Talk to scientists and engineers about their own journey. Ask them open-ended questions about the path that led them to where they are. This can be an interesting and informative discussion for a new scientist or engineer. Ask them about their background and history. What is their educational background? What is their field of expertise? How many years have they been in this field? What is their current position? What work do they do? Ask them about the experiments they've done. What problem were they trying to solve? What gave them the idea to work on this problem? Where do they get ideas and inspiration? How did they go about determining how to resolve questions? What, if any, methods did they use? How important were statistics? How important was the published literature? Did they ever get a null result? Did they see this as a failure? Ask them about other failures in their career. What have they learned from the failures? What impact, if any, did these failures have on their lives and careers? Did anyone criticize them for their efforts? Ask them to describe an example of an iterative experimentation process from their experience and in their field. What is their greatest accomplishment or what are they most proud of?

REFERENCES

Begley, C. G. and L. M. Ellis. 2012. Raise Standards for Preclinical Cancer Research. *Nature* 483:531–533.

Berger, W. 2014. *A More Beautiful Question: The Power of Inquiry to Spark Breakthrough Ideas*. New York: Bloomsbury.

Bode, H., F. Mosteller, J. W. Tukey, and C. Winsor. 1986. The Education of a Scientific Generalist. *The Collected Works of John W. Tukey, Volume III: Philosophy and Principles of Data Analysis: 1949–1964.* ed. L. V. Jones. Pacific Grove, CA: Wadsworth. (The original paper was published in 1949.)

Boring, E. G. 1919. Mathematical vs. Scientific Significance. *Psychological Bulletin* 16:335–339.

Box, G. E. P., W. G. Hunter, and J. S. Hunter. 1978. *Statistics for Experimenters: An Introduction to Design, Data Analysis and Model Building*. New York: John Wiley & Sons.

Brown, L., Editor, 1993. *The New Shorter Oxford English Dictionary on Historical Principles*. 4th Ed. Oxford: Clarendon Press.

Child, J. 1961. *Mastering the Art of French Cooking*. New York: Alfred P. Knopf.

Easton, V. J. and J. H. McColl. 2016. *The Statistics Glossary, v 1.1*. http://www.stats.gla.ac.uk /steps/glossary/.

Economist. 2013. Unreliable Research: Trouble at the Lab. *Economist* October 19.

Ephron, N. 2009. *Julie and Julia*. http://www.sonypictures.com/movies/juliejulia/.

Falin, L. 2013. Correlation vs. Causation: Everyday Einstein: Quick and Dirty Tips for Making Sense of Science. *Scientific American*, October 2. https://www.scientificamerican.com /article/correlation-vs-causation/.

Fanelli, D. 2013. Redefine Misconduct as Distorted Reporting. *Nature* 494(7436):149.

Feynman, R. P. 1965. The development of the space-time view of quantum electrodynamics. Nobel Prize Lecture. http://www.nobelprize.org/nobel_prizes/physics/laureates/1965/feynman-lecture.html.

Feynman, R. P. 1977. *The Feynman Lectures on Physics, Volume 1*. New York: Basic Books.

Goldsmith, B. 2005. *Obsessive Genius: The Inner World of Marie Curie*. New York: W. W. Norton & Company.

Kahneman, D. 2011. *Thinking, Fast and Slow*. New York: Farrar, Straus and Giroux.

Khorasani, F. 2016. Personal communication.

Labouisse, E. C. 1937. *Madame Curie: A Biography*. Translated by Vincent Sheean. New York: Doubleday, Doran & Company, Inc.

Leslie, I. 2014. *Curious: The Desire to Know and Why Your Future Depends on It*. New York: Basic Books.

Levi-Strauss, C. 1983. *The Raw and the Cooked (Mythologiques #1)*. Translated by John Weightman and Doreen Weightman. Chicago: University of Chicago Press.

Nobel. 2016. http://www.nobelmuseum.se.

Randall, L. 2011. *Knocking on Heaven's Door: How Physics and Scientific Thinking Illuminate the Universe and the Modern World*. New York: HarperCollins.

Shermer, M. 2008. Wheatgrass Juice and Folk Medicine. *Scientific American* 299(2):42.

Truran, P. 2013. *Practical Applications of the Philosophy of Science: Thinking about Research*. New York: Springer.

Velickovic, V. 2015. What Everyone Should Know about Statistical Correlation: A Common Analytical Error Hinders Biomedical Research and Misleads the Public. *American Scientist* 103(January–February):26–29. http://www.americanscientist.org/issues/pub/what-everyone-should-know-about-statistical-correlation.

Vigen, T. 2015. *Spurious Correlations*. New York: Hatchette Books. More spurious correlations can be found on his website: http://www.tylervigen.com.

Wainer, H. 2007. The Most Dangerous Equation. *American Scientist* 95:249–256.

Weisberg, R. W. 1993. *Creativity: Beyond the Myth of Genius*. New York: W. H. Freeman and Company.

WTVY. 2016. WTVY broadcasted the Gene Reagan's Noon Farm Report from 1950 to 2005.

2

Eureka! And Other Myths of Discovery

The most exciting phrase to hear in science, the one that heralds new discoveries, is not "Eureka!" but "That's funny..."

Isaac Asimov

In every field, there are myths, and science is no exception. Before we delve any further into problem solving, I hope to dispel several myths about scientific discovery. These myths include the following:

- Great experiments unfold like fairy tales.
- Ideas hit us like lightning bolts.
- Yes, but those scientists were creative geniuses.

There are other myths that we could discuss, but I find that these are some of the more common and dangerous ones, if not physically then to our psyche. The biggest problem with these myths is that they get in the way of many new scientists and engineers and stop others. All the hard work, dedication, dead ends, and failures of real problem solving and experimentation are rarely mentioned.

2.1 FAIRY TALES

In retrospect, a published experiment may look like a perfect story, with the beginning leading inexorably to the ending as a "fairy tale" (see Figure 2.1). When we read article after article in professional journals describing nice, tidy experiments with perfect endings, we tend to assume those

FIGURE 2.1
Fairy tale experimentation.

experiments were set up perfectly. We think we must be doing something wrong. We know that fairy tales end with "…and they lived happily ever after," and we know this isn't reality. Yet we read these perfect, nice, tidy experiments in journals and wonder why our experiments are never that easy.

I wish I could say that experimental iteration was a fairy tale or even a nice clean feedback loop or staircase where each step brings us closer to the next echelon of knowledge, but from what I've seen, this is not the case. There are dead ends and lots of sleepless nights in experimentation. Professor Robert Merton describes the history of ideas as nonlinear and made up of a series of advancing-by-doubling-back (Merton 1965). The story isn't clean nor is it easy to draw the lines between one starting point and another.

All of us who've lived through high school and college labs with less than optimal equipment know the frustration of diligent data collection only to get results that don't make any sense. What follows is an example lab report written by Lucas Kovar (Figure 2.2), while he was a graduate student in physics at the University of Wisconsin (Kovar 2001). I hope it provides a bit of comic relief but also there is a point to including it. Ever feel like doing this exact thing in one of your labs? We laugh because we've all been there. We've all felt that frustration of diligent experimentation, hard work, following all the detailed instructions, and the results are garbage. Maybe you've even run experiment after experiment and cannot seem to impact the results. Not fun, but still a result. In order to complete our lab, we must

Electron Band Structure In Germanium, My Ass

Lucas Kovar
Physics Student

Abstract: The exponential dependence of resistivity on temperature in germanium is found to be a great big lie. My careful theoretical modeling and painstaking experimentation reveal 1) that my equipment is crap, as are all the available texts on the subject and 2) that this whole exercise was a complete waste of my time.

Introduction

Electrons in germanium are confined to well-defined energy bands that are separated by "forbidden regions" of zero charge-carrier density. You can read about it yourself if you want to, although I don't recommend it. You'll have to wade through an obtuse, convoluted discussion about considering an arbitrary number of non-coupled harmonic-oscillator potentials and taking limits and so on. The upshot is that if you heat up a sample of germanium, electrons will jump from a non-conductive energy band to a conductive one, thereby creating a measurable change in resistivity. This relation between temperature and resistivity can be shown to be exponential in certain temperature regimes by waving your hands and chanting "to first order."

Experiment procedure

I sifted through the box of germanium crystals and chose the one that appeared to be the least cracked. Then, I soldered wires onto the crystal in the spots shown in Figure 2.2b of Lab Handout 32. Do you have any idea how hard it is to solder wires to germanium? I'll tell you: real goddamn hard. The solder simply won't stick, and you can forget about getting any of the grad students in the solid state labs to help you out. Once the wires were in place, I attached them as appropriate to the second-rate equipment I scavenged from the back of the lab, none of which worked properly. I soon wised up and swiped replacements

FIGURE 2.2

Lab report written by physics student Lucas Kovar.   (*Continued*)

from the well-stocked research labs. This is how they treat under-grads around here: they give you broken tools and then don't understand why you don't get any results.

In order to control the temperature of the germanium, I attached the crystal to a copper rod, the upper end of which was attached to a heating coil and the lower end of which was dipped in a thermos of liquid nitrogen. Midway through the project, the thermos began leaking. That's right: I pay a cool ten grand a quarter to come here, and yet they can't spare the five bucks to ensure that I have a working thermos.

<u>Result</u>

Check this shit out (Fig. 1). That's bonafide, 100%-real data, my friends. I took it myself over the course of two weeks. And this was not a leisurely two weeks, either; I busted my ass day and night in order to provide you with nothing but the best data possible. Now, let's look a bit more closely at this data, remembering that it is absolutely first-rate. Do you see the exponential dependence? I sure don't. I see a bunch of crap.

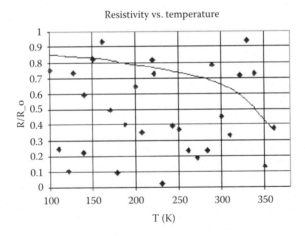

FIG 1: Check this shit out.

Christ, this was such a waste of my time.

Banking on my hopes that whoever grades this will just look at the pictures, I drew an exponential through my noise. I believe the

---

FIGURE 2.2 (CONTINUED)

Lab report written by physics student Lucas Kovar.

(Continued)

apparent legitimacy is enhanced by the fact that I used a complicated computer program to make the fit. I understand this is the same process by which the top quark was discovered.

Conclusion

Going into physics was the biggest mistake of my life. I should've declared CS (computer science). I still wouldn't have any women, but at least I'd be rolling in cash.

FIGURE 2.2 (CONTINUED)
Lab report written by physics student Lucas Kovar.

write up a report and, in the case of Lucas Kovar, hope that the professor has a sense of humor. This feels like the reality of some of our early experimentation. However, when we read professional science and engineering journal articles, they tell a completely different story. Interestingly enough, there are several more parts to this story. Lack of repeatability and reluctance to publish negative results is a part of the story that often goes untold.

A 2013 article published by *The Economist* entitled "Unreliable Research: Trouble at the Lab" broaches the topic of unrepeatable scientific results (Economist 2013). Prior to this, in 2005, Stanford professor of epidemiology and head of METRICS (Meta Research and Innovation Center at Stanford) John Ioannidis presented a paper to the International Congress on Peer Review and Biomedical Publication (Ioannidis 2005). In this work, he reported, "most published research findings are probably false." Additionally, he showed that, statistically, most claimed research findings are false. At the time, only 1 out of every 20 papers reported false-positive results (Ioannidis 2005). Don't get me wrong here: I'm not saying we should question the published results for the band-gap of Germanium. What I am saying is that it's okay to bring a dose of suspicion to newly published research findings. Skepticism is healthy.

There is a lot of pressure in academia to publish. Negative results are not considered interesting by journals. In each publication, authors want to expound on their positive results. Between 1990 and 2007, the publication of negative results across the sciences actually dropped from 30% to 14%, according to Daniele Fanelli while at the University of Edinburgh (Fanelli 2013). Very little information gets published in the sciences related to null or negative results. We don't see these as successes, as opportunities for further learning. We see these as dead ends, as failures.

We become frustrated and look at where to place blame. We assume we did something wrong, or the literature was wrong, or the equipment in our lab was out of calibration. Why? This is a part of experimenting. Determining that something has no effect is an important result. However, we only want to tell people about our experiments when we find something that has an effect or is in close agreement with what others have published. At times, getting results that conflict with published data may be a limitation with our experiment. It may also mean that the published results we are using as our baseline are questionable or were performed in an uncontrolled environment. At other times, the discordant result can be an opening. Harvard University physics professor Lisa Randall explains, "The cracks and discrepancies that might seem too small or obscure for some can be the portal to new concepts and ideas for those who look at the problem in the right way" (Randall 2011).

Null, negative, unclear results happen to any experimenter; the challenge is to maintain a positive attitude when they do. Duke University Professor Dan Ariely tells the story of one of his first experiments at Tel Aviv University (Ariely 2009). Like many of us, he ended up in a psychology class during his first semester. Also, like many of us, this class was impactful in opening up new possibilities for research and areas of study. His professor was instrumental in encouraging him to explore his theories and alternative interpretations to experimental results presented and discussed in class. As an engaged and curious student, Ariely was encouraged by his professor to prepare an "empirical test to distinguish it from the conventional theory." In one case, Ariely developed a hypothesis to test how certain stages of epilepsy developed. He presented the idea and experiment to his professor, who encouraged him to follow through with the experiment. Ariely and a buddy spent the next several months operating on rats only to find out his theory was wrong. He describes his experience: "I was able to learn something about my theory, after all, and even though the theory was wrong, it was good to know this with high certainty." Ariely, the student, was not discouraged by his experience but now intrigued and armed with a new understanding "that science provides the tools and opportunities to examine anything I found interesting" (Ariely 2009).

Although many of us enjoy linear thinking, experimental discoveries aren't linear. Scientific progress proceeds in fits and starts. Imagine assembling a jigsaw puzzle. If we get stuck in one area, we move to another area. Consider a Su-Do-Ku puzzle. We work on one area until we aren't making any progress and then we move to another area. Solving problems in science is similar. The path to get to the solution is not known, and in all likelihood, there are multiple ways to reach a discovery. As Walter Isaacson writes the story of the history of Silicon Valley, ideas come from many sources, converging and diverging at the present moment (Isaacson 2014).

Ideally, scientific research is a process of guided learning. According to Dr. Kevin Ashton in his book *How to Fly a Horse*, "Imagination needs iteration. New things do not flow finished into the world. Ideas that seem powerful in the privacy of our head teeter weakly when we set them on our desk. But every beginning is beautiful. The virtue of the first sketch is that it breaks the blank page. It is the spark of life in the swamp. Its quality is not important. The only bad draft is the one we do not write" (Ashton 2015). The object of the methods and tools presented herein is to make the process of discovery as efficient as possible. However, we may feel as if we are caught in the scary maze of experimentation, as in Figure 2.3. We may learn in an iterative manner, but the act of creation is more like wading through a maze where, at each step, we stand on the shoulders of someone who came before us (Ashton 2015). Astrophysicist Mario Livio describes the evolution of scientific progress in a description of the theory

FIGURE 2.3
Experimentation may feel like a very scary maze at times.

of evolution. "The story of evolution is therefore not a simple narrative leading from myths to knowledge but a collection of diversions, blunders and winding paths. Eventually, all of these intertwined threads converged into one conclusion…" (Livio 2013).

An initial hypothesis leads by a process of deduction to certain necessary consequences that may be compared with data. When the consequences and data fail to agree, the discrepancy can lead, by a process of induction, to modification of the hypothesis. A second cycle of the iteration is thus initiated. The consequences of the modified hypothesis are worked out and again compared with data that in turn leads to further modification and gain of knowledge. We have a hypothesis or theory, H_1, which leads to modified hypothesis or theory, H_2; then H_2 leads to another idea, H_3, and so on. This strategy has been described as a feedback loop as if we were in a computer program (Box et al. 1978). In reality, just like the nice clean feedback loop and the fairy tale, the experience of experimenting can feel much more like a maze. As Professor Feynman describes, "We have a habit in writing articles published in scientific journals to make the work as finished as possible, to cover up all the tracks, to not worry about the blind alleys or describe how you had to the wrong idea first, and so on. So there isn't any place to publish, in a dignified manner, what you actually did in order to get to do the work" (Feynman 1965). In the Nobel Museum, experimentation and problem solving are described as a "multifaceted process. It alternates between thinking and practical work, between intensive reappraisal, repetitive routines and anticipation" (Nobel 2016).

2.2 LIGHTNING BOLTS

Dr. Ashton opens his book *How to Fly a Horse* with a reprint from 1815 *General Music Journal* that was rumored to have been written by Mozart about his creative process. "When I proceed to write down my ideas the committing to paper is done quickly enough, for everything is, as I said before, already finished; and it rarely differs on paper from what it was in my imagination" (Ashton 2015). The evidence used by many of Mozart's effortless compositional creations are the many perfect manuscripts. There are no fixed mistakes. Although it has continued to be referenced by many authors, this document is a forgery. Mozart's widow kept his manuscripts but stated in a letter that she had discarded the "unusable autographs" before selling the rest (Weisberg 1993). This creation myth is not even close to resembling the real creative struggle that Mozart went through. Yes, he was gifted in music, but his work was not magical. There was no dream or lightning bolt that struck him and delivered complete symphonies. What was his secret? It was work. He wrote and rewrote scores.

We often describe a discovery as a light bulb coming on. I love the following bit of trivia on the history of the link between the light bulb and a bright idea. In 1919, before audio was integrated into films, there was a cartoon character named Felix. Felix the cat was the brainchild of artist Otto Messmer and producer Pat Sullivan. Felix used the appearance of symbols and numbers in the film as objects of opportunity. In Felix's films, light bulbs would appear above his head when he had an idea. This symbol has long outlived its originator, yet when we become aware of something new, this light bulb comes on for us. We could think of everything that we don't know as dark space. As we experiment and try new things, we work for light bulbs to illuminate these great unknowns.

Progress has an iterative nature. "Make small changes, small changes right where you are. Large changes occur in tiny increments. Small actions lead to larger increments in our creative lives. Take one small daily action instead of indulging in the big questions. Creativity requires action ... We prefer the low-grade pain of occasional heart stopping to the drudgery of small and simple daily steps in the right direction" (Cameron 1992). Double Nobel Prize winner Marie Curie wrote, "A great discovery does not issue from a scientist's brain ready-made, like Minerva springing fully armed from Jupiter's head; it is the fruit of an accumulation of preliminary work" (Goldsmith 2005). Science progresses incrementally. Professor Randall, in *Knocking on Heaven's Door*, wrote, "That's how science works. People have ideas, work them out roughly, and then they or others go back and check the details. The fact that the initial idea had to be modified after further scrutiny is not a mark of ineptitude—it's a sign that science is difficult and progress is often incremental" (Randall 2011). A friend of mine was famous for showing up at engineering review meetings and asking us "How do you move a ten ton weight down the street?" The answer is "one inch at a time." This is the really tricky part: keeping our eyes on the goal but moving inches forward every day until we arrive at our destination.

Even incorrect theories and explanations of results can potentially be viewed as progress, one inch down an incorrect path we learn there is no need to proceed any further. "Science progresses not in a straight line from A to B but in a zigzag path shaped by critical reevaluation and fault-finding interaction. The continuous evaluation provided by the scientific establishment ... creates the checks and balances that keep scientists from straying too far in the wrong direction" (Livio 2013). Astrophysicist Fred Hoyle was the first to propose a steady-state universe. This was later shown to be incorrect by Edward Hubble and Georges Lemaitre. However, Hoyle's "wrong" theories were a catalyst for others, adding fuel to the fire, and generating fresh thinking in new fields of study. "... There is no greater myth worth dispelling in the history of innovation than the idea that progress happens in a straight line" (Berkun 2010). Another description from a wall in the Nobel Museum in Stockholm, experiments and problems solving "... run like a golden thread through history ... usually not in direct routes but with twists and turns, offshoots and cross connections ... their results present a multifaceted and fluid picture—a constant interplay of tools, ideas and discoveries" (Nobel 2016).

It is only through the thoughtful and persistent amassing of knowledge that has been gathered and worked over, bread-kneaded in our minds, so to speak, that eureka moments occur (Duckworth 2016, Leslie 2014, Oakley 2014). "Fortune favors the prepared mind," wrote Louis Pasteur. He knew, as did all of the great scientists I've mentioned so far (who are regularly called geniuses), that his accomplishments came out of, wait for it, W–O–R–K. For both Isaac Newton and Mozart, the secret was work. These great scientists continued to work and struggle, almost obsessively, on their subject.

In his book *Curious: The Desire to Know and Why Your Future Depends on It*, Ian Leslie relates an example of how preparation is so essential for an epiphany moment with an anecdote from French mathematician and engineer Henri Poincaré. Poincaré had been wrestling with a pure mathematics problem for a number of months. He was called to visit a mining site. In Poincaré's words, "As I put my foot on the step the idea came to me, without anything in my former thoughts seeming to have paved the way for it, that the transformations I had used to define the Fuchsian functions were identical with those of the non-Euclidean geometry" (Leslie 2014). Ideas come to us after we mull over problems, steep in confusion, bump into multiple dead ends, and then finally, maybe, the solution comes to us.

Although this is an area of study that dates back many years, recently, neuroscientists have linked rapid eye movement sleep with these "connections between different associate networks of knowledge." When our brains are the most free and relaxed and our dreams the most vivid, accumulated knowledge makes itself available for arrangements that our conscious mind would never consider. It is as if the conscious mind is able to release the grip on our thoughts and the information in our brains is able to move around freely like strangers at a dinner party (Leslie 2014).

2.3 GENIUSES

Griffins, ghouls, gnomes, giants, and geniuses are all mythical creatures. I apologize for the alliteration, but this particular myth has stopped so many people from pursuing science and engineering. I wanted the alliteration to provide something that would be easy to remember. There are several variations of this particular myth that we say to ourselves which stops us: either "I'm not a genius" or "I'm not creative" or "I'm not naturally talented." These statements serve to keep many people from ever getting started with anything, especially the sciences.

Over the years, there have been many scientists who have attempted to prove that certain types (races, genders, ethnicities, etc.) of people are geniuses or inherently had more aptitude to learn than others. Repeatedly, these experiments have failed and in many cases missed children who have gone on to win Nobel Prizes. One such example was Stanford University Professor Lewis Terman, developer of the Stanford-Binet IQ test, who identified 1500 children of exceptional abilities (whose geniuses were called "Termites" and who had IQs on average of 151) and rejected 168,000 others as ordinary. The Termites were studied for more than 35 years. Some of them did great things but others went on to live ordinary lives. (His work is actually very controversial and flawed, I'm only referencing it here as an example of one experiment that dispels the myth of genius.) The real story is the rejected majority (Ashton 2015). Included in this rejected majority were Physics Nobel Laureates William Shockley and Luis Alvarez. Shockley's Nobel Prize was for the coinvention of the transistor and Alvarez won for proposing that an asteroid may have crashed to Earth and killed the dinosaurs. How could they have been missed by a genius test?

Unlike when Terman developed the IQ test, we now know that IQ can change, grow, and develop. With increased exposure, new neuronal pathways continue to form. We also now know that IQ is not indicative of future accomplishments, contributions, or potential. There it is—I am not a genius, you are not a genius, but, on the bright side, neither is that guy who received an A on the Algebra II exam or the first chair violinist or the chess club president. No one starts out extraordinary. The first time we draw a picture or pick up a musical instrument, it is unlikely to be good. Recall the words of Thomas Edison: "Genius is one percent inspiration, ninety-nine percent perspiration." If we do the math, that doesn't leave much room for IQ. Although Edison may not have worked this math out literally, he was on to something. We are only recently beginning to unravel the complex connections that have some people achieve great things. Edison was on to it. As University of Pennsylvania's professor of psychology Angela Duckworth has shown in her research, the secret to success is not talent or genius but a combination of persistence and passion which she calls grit (Duckworth 2016). Temple University psychology professor Robert Weisberg has found in his studies of problem solving that underlying creativity is cognitive thought with "very high degrees of persistence and motivation." (I will repeat this exact statement later in the book.) Professor Weisberg showed that it takes ordinary thought processes to produce novel behavior (Weisberg 1993).

2.4 KEY TAKEAWAYS

Every scientist or engineer who has worked in a lab or performed an experiment has been affected to some extent by one or all of the myths discussed in this chapter. Even today, we continue to perpetuate these myths. Read almost every technical journal and it appears that the experiments were magical. They read like the fairy tales of our childhoods. The true story, all the dead ends, all the stumbles and falls, are omitted. It can be frustrating, but don't be discouraged. We don't know exactly where ideas come from. We know that "lightning strikes" of ideas are rare, if they have ever really occurred. What appears to have been an "Aha" or "Eureka" moment was really the result of a lot of hard work, stewing over ideas and concepts and taking lots of erroneous dead ends. Similarly with geniuses and "naturals," there are those who succeed, but those successes are a result of many hours of practice and hard work. As we learn more about human behavior, we realize that passion and perseverance are really the keys to success. The results from our experimentation WILL BE consistent with our experimental setup. Therefore, we need to make sure that our experimental setup is as good as we can make it. In the upcoming chapters, we will begin to examine sources of variation which, gone unchecked, can affect our experiment. This will allow us to truly explore the effect(s) of interest. First, however, let's cover an important and critical topic: communication.

P.S. In the face of discouragement, when we aren't the top in the class or bad things happen, that is when our grittiness needs to kick in. Think of an experiment that failed or didn't go the way it should have. Do a postmortem on this failure. Can you identify why it didn't work? Try to come up with a short list of maybe 10 things that could possibly have changed the outcome of the experiment.

REFERENCES

Ariely, D. 2009. *Predictably Irrational: The Hidden Forces That Shape Our Decisions.* New York: HarperCollins.

Ashton, K. 2015. *How to Fly a Horse: The Secret History of Creation, Invention, and Discovery.* New York: Doubleday/Random House.

Berkun, S. 2010. *The Myths of Innovation.* Sebastopol, CA: O'Reilly Media.

Box, G. E. P., W. G. Hunter and J. S. Hunter. 1978. *Statistics for Experimenters: An Introduction to Design, Data Analysis and Model Building.* New York: John Wiley & Sons.

Cameron, J. 1992. *The Artist Way: A Spiritual Path to Higher Creativity.* New York: Jeremy P. Tarcher/Putnam.

Duckworth, A. 2016. *Grit: The Power of Passion and Perseverance.* New York: Scribner/Simon & Schuster.

Economist. 2013. Unreliable Research: Trouble at the Lab. *The Economist* October 19.

Fanelli, D. 2013. Redefine Misconduct as Distorted Reporting. *Nature* 494(7436):149.

Feynman, R. P. 1965. The development of the space-time view of quantum electrodynamics. Novel Prize Lecture. http://www.nobelprize.org/nobel_prizes/physics/laureates/1965/feynman-lecture.html.

Goldsmith, B. 2005. *Obsessive Genius: The Inner World of Marie Curie.* New York: W. W. Norton & Company.

Ioannidis, J. P. A. 2005. Why Most Published Research Findings Are False. *PLoS Medicine* 2(8):e124.

Isaacson, W. 2014. *The Innovators: How a Group of Hackers, Geniuses, and Geeks Created the Digital Revolution.* New York: Simon & Schuster.

Kovar, L. 2001. Germanium Band Gap, My Ass. *Annals of Improbable Research* 7(3). www.improbable.com/magazine. The complete paper can also be found at http://pages.cs.wisc.edu/~kovar/hall.html.

Leslie, I. 2014. *Curious: The Desire to Know and Why Your Future Depends on It.* New York: Basic Books.

Livio, M. 2013. *Brilliant Blunders: From Darwin to Einstein—Colossal Mistakes by Great Scientists That Changed Our Understanding of Life and the Universe*. New York: Simon & Schuster.

Merton, R. K. 1965. *On the Shoulders of Giants: A Shandean Postscript*. New York: The Free Press.

Nobel. 2016. http://www.nobelmuseum.se.

Oakley, B. 2014. *A Mind for Numbers: How to Excel at Math and Science (Even if You Flunked Algebra)*. New York: Jeremy P. Tarcher/Penguin.

Randall, L. 2011. *Knocking on Heaven's Door: How Physics and Scientific Thinking Illuminate the Universe and the Modern World*. New York: HarperCollins.

Weisberg, R. W. 1993. *Creativity: Beyond the Myth of Genius*. New York: W. H. Freeman and Company.

3

Experimenting with Storytelling

Data alone won't get you a standing ovation.

Carmine Gallo

An engineer or scientist who operates in a vacuum may not need to communicate, but the rest of us will need to. In school, effective communication allows professors and instructors to verify that we understand the materials of the course. In the world of industry, government, or other academia roles (our jobs), our engineering managers, marketing team, and executive staff need to understand clearly and concisely what we discover and learn from our work. Communication is a whole package: the words we use, the tables we create, and the graphics we generate. The words, phrases, terms, and text are critical to clear communication. Visual displays/pictorial representations, whether they are photographs, micrographs, graphs, or charts, are important tools in the effort to communicate data. To accomplish the goal of communication in such a manner that inspires confidence, the data must be clearly, concisely, and accurately represented. The complexity of scientific data often requires the use of all forms of communication—words and graphics. The more unified our words and graphics are, the more effective and easier the communication. Allowing our experiments to unfold as with stories can engage, captivate, and even delight our audiences.

There are a variety of forms we can utilize to communicate our data. The tools we use to describe our experiments, observations, and the results are (1) words or language, (2) visual displays, and/or (3) numbers displayed in tables or graphs. In this chapter, we'll look at each of these separately. Often, the final or eventual task of any data collection and analysis is to summarize the data and communicate the results from the experiment in such a way that a follow-up decision can be made. Whether verbally or in written form, scientists must use words, graphics, tables, and statistical summaries to effectively communicate their experiments and the findings. As engineers and scientists (whether professional or amateur), we need to be fluent in all of these types of communication.

3.1 THE SECRETS OF SCIENCE

It wasn't until around the time of the Royal Society that it became commonplace to actually share scientific findings. Beforehand, there was a code of secrecy in the ancient world among scientists. Mathematician Girolamo Cardano wrote to a colleague, "I swear to you by God's Holy Gospels and as a true man of honor, not only never to publish your discoveries, if you teach me them, but I also promise you, and I pledge my faith as a true Christian, to note them in code, so that after my death no one will be able to understand them." According to legend, Pythagoras banished one of his followers, Hippasus, for "telling men who were not worthy" a dreadful mathematical secret. Hippasus had shared with an "outsider the discovery that certain numbers couldn't be written precisely." To the Greeks, this was a nightmare (Dolnick 2011). The Royal Society ended this period of secrecy, and science and scientific experiments began to be shared and made accessible.

Isaac Newton was one of the first scientists to share his experiments. Newton invented calculus in an effort to solve physics problems. However, he had no intention of publishing or sharing his findings. Some 30 years after Newton's development of calculus, Edmund Halley convinced him to share his results with the world. Halley funded and micromanaged each detail of the publication (Livio 2013). In the foreword to *Principia*, Halley wrote (Newton 1999):

> The things that so often vexed the minds of ancient philosophers
> And fruitlessly disturb the schools with noisy debate
> We see right before our eyes, since mathematics drives away the cloud.

Independent of why we are investigating, at some point, we most likely will need to communicate how we arrived at our findings. Whether we are in high school, a university student, attending a conference, or part of the workforce, we need to communicate effectively in order to have our audience take us seriously. I want to stress the critical nature of this part of experimentation by ensuring that communication is at the forefront of discussion.

Developing dexterity with all forms of communication gives our intended audience confidence not only in the data but also in us as engineers and/or scientists. The form of communication may vary depending on our situation but, "the ability to communicate is a very important skill." Statistician Dr. Fred Khorasani labels communication dexterity as one of the basic skills in engineering. The other basic skills required to do our experiments may change over time. For example, many years ago (before my time), using a slide rule was a basic skill for engineers and scientists, but today navigating specialized software packages is a basic skill. Dr. Khorasani writes, "People with basic skills are much more effective in investigation or in problem solving" (Khorasani 2016). The stronger our basic skill set is, the more we will be able to do and the more effective we will be in the long run.

In this chapter, we will cover the language of science as well as when and how to use graphics and tables. The other communication tool, statistical summaries, will not be specifically addressed in this chapter. Experimentation, measurement, and statistics form a holy trinity. The measurements involved in the early sciences—first astronomy, then experimental physics—put increased pressure on mathematicians to understand and quantify random error. These needs drove the development of statistics. Statistics "provides a set of tools for the interpretation of data that arise from observation and experimentation. … But statistics also provides tools to address real-world issues, such as the effectiveness of drugs or the popularity of politicians, so a proper understanding of statistical reasoning is as useful in everyday life as it is in science" (Mlodinow 2008). We will cover statistical summaries throughout the remainder of the book as this may be the least familiar communication tool.

3.2 THE LANGUAGE OF SCIENCE

The act of skillful writing schools its author in ways of explaining structure and significance, of explaining ideas. Which is just what you need to do good science.

Roald Hoffman

I introduce this idea of language because it is through language that we communicate the results of our experimentation. Effectively communicating our findings with language gives others access to our work and thereby to us as scientists and engineers. The language of science establishes a foundation for us to build and share concepts, ideas, and results. Sharing fact and rattling off data and definitions are not what science and engineering are all about. When we reach out with "experimentally-centered explanations about how certain concepts were discovered", we want to connect with our audience by giving them the "who, what, when and how" of our experimentation (Williams 2013). Let's take a look at early attempts at communication in science.

To illustrate the point of how critical language is, let's walk through the story of Gerolamo Cardano, an Italian medical doctor born in 1501. Cardano wasn't born into a wealthy family and decided to play games of chance to pay his way through medical school. He developed a system with games that required skill and a system with those that were pure chance. He understood odds and was therefore quite successful. At 19 years old, he had saved up enough money for his education. Cardano decided to write down his theory of gambling (Mlodinow 2008).

Let's back up a moment and look at what was going on at that time in mathematics. The symbols + and − had not been introduced yet for what they mean today. (They were introduced by the Germans to indicate overages or underages in chests.) The abbreviations p and m were used for plus and minus in the fifteenth century. Around AD 700, the Hindus had introduced the base 10 positional notation (the standard), which reached Europe. However, the equal sign wasn't introduced until 1557, and the sign for multiplication wasn't introduced until the seventeenth century. The ideas of Descartes and coordinates that unify algebra and geometry were more than a century later than Cardano. It was with these challenges that Cardano wrote *The Book on Games of Chance*. Cardano wrote the first book on probability, but his notation was such that no one could decipher it, so probability wasn't understood until much later. He hadn't established a notation or linguistic abstraction that would allow for easy communication; therefore, his contemporaries couldn't understand his ideas of randomness and chance and thought he was a demon or cursed because of his use of probability theory in gambling. By the way, the book wasn't published until many years after it was written, roughly 100 years after Cardano's death (Mlodinow 2008).

Galileo faced similar trials as Cardano. Galileo wanted to describe the world and his observations of the world in mathematical proportions. Because the algebraic symbols that today we commonly use to communicate motion had not been established, Galileo wrote his proofs and theorems in dense text accompanied by detailed letter labeling and descriptions (Sobel 1999). Isaac Newton also faced Cardano's communication challenges when he began his study of velocity. Newton found that in order to communicate his scientific findings, he first had to develop calculus (Dolnick 2011).

We fortunately do not face the same challenges as Cardano or Galileo. As scientists and engineers, we have an established language and set of universal notational symbols in our fields and mathematical notations that serve us. Although discoveries are happening every day that add new linguistic abstractions to the lexicon, we now have access to and facility with the language of science. The language of science has a specific conversational domain, which creates a network of terms. These terms, which include our symbols, notations, and equations, are agreed upon by the scientific community and allow us to comprehend and interact with one another more efficiently. If we use standard notation for Newton's Law, $F = ma$, when communicating with other scientists, we need not explain that the symbols F represents force, m represents mass, and a represents acceleration. This is common knowledge among all science and engineering students. However, if for some reason we expressed Newton's Law as $G = nb$, this form of the equation would require an explanation to clarify that G is used to represent force, n is mass, and b is acceleration.

Although we now have more access to the language, symbols, etc., communication of science and engineering phenomena remains an issue centuries later. Dr. Edward Deming devoted a chapter in his 1982 book *Out of the Crisis* to this issue of clear, well-defined communication. "The only communicable meaning of any word, prescription, instruction, specification, measure, attribute, regulation, law, system, edict is the record of what happened on application of a specified operation or test" (Deming 1982). The meaning of any communication must have the same definition independent of the reader or person attempting to use the communiqué. In the classroom, we talk about ideal situations. To describe something as round is to have it meet the Euclidean requirement of being equidistant from the center. When we begin to apply "round" in the laboratory, it quickly becomes clear that this definition is just formal logic or a concept.

In practical applications, we need to define "round enough" for our laboratory. Struggling with communication early in our careers will allow growth and development of an incredibly valued muscle in the future.

When a new field of science is developed, a new conversational domain is required that allows for the expert practice of this new field. The mastery of these specialized terms, notations, symbols, and tools provides access to this new scientific field, which might not otherwise be available. We use this specialized language when communicating in our area to give others a feel for this new field and provide this access for others. We use this language to shape the way this area of study shows up for ourselves and others.

Many of us recall how difficult the new language of physics, biology, chemistry, or mathematics was at first. We spent a lot of time learning and digesting new ideas, words, and definitions. Some of us may have even spent our whole high school and undergraduate careers wrestling with the definitions in these fields. Even though at first it may seem awkward or difficult, if we stick with it, eventually, these concepts and ideas become accessible to us, almost commonplace. We become comfortable with the ideas represented by these terms.

The language that we use to communicate science, experiments, and/or research is the same language used by the dullest author and the most captivating author. We want our writing to be read, possibly multiple times. The more enjoyable our writing is, the more appealing the work will be to readers. In light of this, the balance of the chapter covers the tools we use to communicate from storytelling to presentations. Equations, numbers, diagrams, charts, graphs, micrographs, photographs, and words are all a part of the language of science that we use to share experimentation. "It is all information after all. ... Most techniques for displaying evidence are inherently multimodal, bringing verbal, visual, and quantitative elements together" (Tufte 2001). The concise and accurate representation of data to communicate information is both art and science. Words and pictures can be used to summarize details into information. Graphics, tables, and text allow us to share information gleaned from raw data. Statistics provide a tool for describing large data sets in one or two numbers and we'll be looking at statistical summaries in later chapters. In this chapter, we will review some commonly used communication tools: storytelling with words and graphical techniques, both of which can be helpful in the problem-solving setup and then useful in the communication of the experimental setup and presentation of the data. The words and graphical devices we discuss can be used together or have stand-alone applications. However, they are best used together. First, let's take a quick look at the most common venues for presenting scientific and engineering results.

There are four primary ways we share/communicate information in the sciences:

1. Lab reports or dissertations in academia or internal memos in business,
2. Journal articles,
3. Posters, and
4. Talks or oral presentations.

Table 3.1 compares typical styles and audiences of different venues. There are some commonalities about these venues—we are either writing or speaking. Although the intended audience may be large or small, we primarily present to our colleagues, an audience who is somewhat familiar with our work.

TABLE 3.1

Common Venues for Sharing Engineering and Scientific Results

Venue	Format	Audience	Audience Size	Features
Lab report, dissertation, internal memo	Written	Professors, teaching assistants, small team	Small group	Less formal, more data heavy
Journal article	Written	Other professionals some very familiar with topic	Larger audience dependent on journal circulation	Formal, language should be concise and clean
Poster	Oral with aids	May be familiar with topic	Small groups at a single time but large groups overall	Less formal but not casual; free form Q&A with audience
Talk/speech	Oral, may include aids but not essential	Typically familiar with topic	Tends to be larger audience	Formal, more structure Q&A

Although we may not be interested in pursuing an academic career, writing a technical paper or presenting a poster is an excellent experience, in particular, if the problem is one that has sparked our interest and we can share our personal joy of discovery. There are opportunities for sharing and publishing our work even if we are fairly new to problem solving and experimentation. The *Journal of Young Investigators* has an international audience and is a respected venue for publishing our early work. Several universities have their own undergraduate journals which offer platforms for publishing experiments. The *Journal of Young Investigators* was established to provide just such a home for the work of new engineers and scientists. Peter Kalugin, a Johns Hopkins University student and editor-in-chief of the *Journal of Young Investigators*, commented, "Too often, people ignore that a key part of being a scientific leader is being a communicator" (Kim 2015). Professional student organizations will often attend annual meetings with other student organizations that provide opportunities for undergraduates to present their work. I can't stress enough how important participating in these opportunities can be. Even as the presenter, we learn so much about our topic by sharing it with others. In an article by Yoo Jung Kim entitled "How Undergraduate Journals Foster Scientific Communication," Kim explains that undergraduate journals help students summarize complex topics and hone their writings skills while getting an introduction to the world of academic publications (Kim 2015). These opportunities to participate in the scientific academy provide opportunities that allow less experienced problem solvers to "learn to write and talk comfortably in a scientific context," adds Kalugin (Kim 2015).

To aid in selecting the right communication tool, we need to answer some questions early on. What are we attempting to analyze or compare? Are we attempting to show a pattern/trend or make a prediction? Once the data are analyzed, the bar for communicating that information is in the answer to the following question. What is the most *efficient* display of *meaningful* and *unambiguous* data (Klass 2012, Tufte 2006)? The answer to these questions will help us determine how best to communicate our results. The words in italics are the bar that we need to rise to meet. They are the standards we want to hold ourselves to. If we do all of this and it still isn't clear which is the best tool to use for communicating, we can try different approaches to see what works best or use complementary techniques. Words, tables, or graphics can be used in all typical formats of presentation. Deciding when to use each one is the tricky part. The all-encompassing definition of communicable experimental information includes words, drawing, micrographs, photographs, two-dimensional (2D) and 3D graphs and charts, equations, etc. The more time we spend upfront deciding how to communicate our data and clarifying what it means, the easier it will be for the audience to understand and grasp what we have discovered.

3.3 STORYTELLING WITH DATA

The skill of writing is to create a context in which other people can think.

Edwin Schlossberg

It appears to be a source of pride for scientists and engineers that our writing be dull. "Most academics find getting the initial ideas the most enjoyable part of research and conducting actual research is almost as much fun. But few enjoy the writing, and it shows. To call academic writing dull is giving it too much credit. Yet to many, dull writing is a badge of honor. To write with flair signals that you don't take your work seriously and readers shouldn't either," writes University of Chicago Business School Professor Richard Thaler, the father of the field of behavioral economics (Thaler 2015). We can fill our papers or talks with lots of "technomumble-jumble"; however, the risk we run with this "showboating" is that we lose our audience (Williams 2013). We fail in our attempts to have them see our discovery, our learnings. What if we could grasp something from every paper we read? What if we enjoyed journals written by our peers? What if our writing inspired and captivated others? Let's look at one way that this may be possible. Physics Nobel Prize Laureate and former California Institute of Technology Physics Professor Richard Feynman is a classic exception to the dull, dry, boring academic writing. A reader need only read one chapter of his books to understand the absolute pleasure Professor Feynman had with discovery. Professor Feynman was a storyteller.

Storytelling may seem like an odd topic for a book on experimentation, even if the chapter topic is communication. Just so there is no confusion, I'm not talking about fictional storytelling. However, I am choosing storytelling very deliberately and intentionally. As human beings, we love stories. It is one of the few truly universal traits that we share across cultures and throughout history (Hsu 2008).

We have a roughly 5000-year-old oral tradition (Houston 2008) and a 2700-year history from the first cave paintings. Storytelling, either spoken, written, or drawn, has spanned this history. Storytelling has given us fairytales, mysteries, fiction, and nonfiction. "Storytelling is both ancient and deeply human. It is a shared treasure between science and the arts and humanities," writes Cornell Professor Roald Hoffman, who received a Noble Prize for chemistry in 1981. Professor Hoffman expresses it simply, "In the papers I read and write, I feel stories unfold before me. I react to them emotionally. I sense narrative devices in these articles and lectures, employed both spontaneously and purposefully" (Hoffman 2014).

Narrative is a form of story structure that uses a "series of causally linked events that unfold over time" (Hsu 2008). Our brains are wired to look for a cause-and-effect relationship. Andrew Stanton, Pixar screenwriter and director, explained in his TED talk, "We're born problem solvers. We're compelled to deduce and to deduct, because that's what we do in real life. It's this well-organized absence of information that draws us in … and it's like a magnet. We can't stop ourselves from wanting to complete the sentence and fill it in." Stanton continues, "Storytelling … [is] knowing your punchline, your ending, knowing that everything you're saying from the first sentence to the last, is leading to a singular goal, and ideally confirming some truth that deepens our understandings of who we are as human beings" (Stanton 2012).

In sharing our experiments and findings, whether in writing or orally, we can envision ourselves as narrators of our experimental discovery. We are "the spinner of theories, the sequencer of steps in a chemical reaction" (Hoffman 2014). The communication about our experiment can be more than just listing of facts. We can discuss how concepts were discovered, and this is truly much more representative of how we actually approach problems. This approach creates some mystery, which actively engages readers and listeners in trying to determine what is going to happen. It is this approach that will inspire thought and motivate interest in our work. "It's how we discuss it, interact with it, and learn about it that makes a subject accessible," writes University of California graduate student Holly Williams (2013).

The story of any experiment is filled with mystery and suspense, which are two key ingredients in fiction works. As investigators, we are unlocking or unraveling the mysteries of nature. We consider all that we've been taught, all that others have discovered, all that we've read about, and we begin to ask questions. We develop hypotheses. These hypotheses are in essence stories about what might be happening. We creatively struggle with how best to solve our mystery. We consider different options and approaches. We have to make choices and decisions that may impact the results. We perform the experiment and then it is time for interpretation of a set data. What do the results mean? What impact will these findings have? What are the broader implications? With the new experimental results in hand, we build "a face of reality" (Hoffman 2014). The world is then seen in a different light by those who read or learn of this work.

When communicating scientific work, we are of two minds. We are the author of the story and one of the main characters. "The protagonists are the investigators of nature," advises Professor Hoffman. We must be in the story as observer and interpreter. We must grapple with performing our observations and measurements. At the same time, we are asked to interpret these findings. We are asked to frame the story for the reader or listener. "Carefully done measurements of observables are an essential ingredient of science, against which theories must be measured. They constitute facts, some will say. Well, facts are mute. One needs to situate the facts, or interpret them. To weave them into nothing else but a narrative" (Hoffman 2014).

Good stories teach us. We learn from stories. There is a wonderful rich research area in neuroscience and psychology studying the effects of stories on our brains. Researchers are looking at which areas of our brains are activated while reading and listening. They are looking at how our brains couple to and mirror as we listen. We are discovering how important stories are to learning and developing relationships within a social world (Hasson et al. 2012). Stories "cross the barriers of time, past, present and future, and allow us to experience the similarities between ourselves and through others, real and imagined" (Stanton 2012).

We've seen that Galileo and Newton dismantled Aristotle's armchair scientific "sit back and think about it" philosophy. However, in his role as a philosopher, and given what it takes to effectively communicate, Aristotle was on to something. He proposed that persuasion (effective communication) had three components: ethos, logos, and pathos. These are Greek words meaning character, logic, and experience. We can think of ethos or character as our expertise or reputation. This makes us credible subject matter experts. Logos is all the data, statistics, and logical arguments we use to back-up our claims. Finally, there is pathos, the experience of our invention or discovery. The nonprofit Technology, Entertainment and Design (TED) Ideas Worth Sharing has set a new presentation bar (Anderson 2016). TED brings together people from all walks of life with the goal of changing the world through the sharing of ideas. Carmine Gallo, author and communication coach, has analyzed TED presentations from some of the greatest speakers in the world. Gallo found that ethos and logos only account for less than half of the presentation. More than half of the presentation was pathos (Gallo 2014).

When we present the results of our experiments, we should strive to connect with our audience using the story of our investigation. Professors Feynman and Hoffman perfected the art of storytelling. Their books and lectures are filled with stories that allow them to connect with everyone in the audience. University of Houston Graduate College of Social Work Professor and #1 New York Times best-selling author, Brene Brown, suggests that storytelling is a means of accomplishing pathos. Although she is a world-renowned expert in social work, Professor Brown gladly accepts the title "Storyteller" when she presents. She tells stories with her data. "Maybe stories are just data with soul," she suggests, "… we're all storytellers" (Gallo 2014).

Gallo writes, "Researchers have discovered that our brains are more active when we hear stories. A wordy PowerPoint slide with bullet points activates the language-processing center of the brain, where we turn words into meaning. Stories do much more, using the whole brain and activating language, sensory, visual, and motor areas" (Gallo 2014). Dissertations and journal articles, regardless of the field, are still for the most part expositions, straightforward explanations with lists of facts and figures. Although most academic journals have a formal outline that must be followed for an article to be published, as engineers and scientist, we have the option to communicate our work in such a way that both our story as well as nature's story comes through.

3.4 STORYTELLING WITH GRAPHICS

We can use a thousand words to describe experimental results or we can show a picture or we can use a combination of the two. The complexity of the data we generate in our experiments can require multiple forms of presentation. The data we collect are useful and impactful only if our intended audience comprehends our data and the implications of that data. Visual displays are intended to enrich and enhance the text (the story of the experiment). With effective graphical tools, we may be able to reduce the thousand words to something more manageable.

Reports, papers, articles, or presentations use graphics and words together to describe the experimental investigation and the findings. Graphs are a great tool for presenting large data sets concisely and in a coherent manner. In typical scientific reports (journal articles, lab reports, etc.), visual displays account for up to 50% of real estate on the page and occasionally more. The graphics we use should enrich and supplement the text, equations, tables, and statistical summaries. An excellent graphic will summarize and display complex ideas in a clear, precise, and efficient manner.

When choosing how best to display data, it is important that there be a clear purpose for the graphic. The purpose could be description, exploration, comparison, tabulation, or decoration. The intent of the graph should be clear. Good graphical displays show data in a completely self-explanatory manner. "The greatest value of a picture is when it forces us to notice what we never expected to see" (Tukey 1977). The focus of the graph should be on the data, not on the "methodology, graphic design, the technology of graphic production, or something else" and "not how perfectly stylish the pages look" (Tufte 2001, 2006). Good graphics reveal data and don't distract from or distort the results.

Graphics can be more precise and revealing than regular statistical computations. As Professor Tufte wrote, "the essential test of text/image relations is how well they assist in understanding of the content." He continues, "Evidence is evidence, whether words, numbers, images, diagrams, still or moving. It is all information after all. For readers and viewers, the intellectual task remains constant regardless of the particular mode of evidence: to understand and to reason about the materials at hand and to appraise their quality, relevance, and integrity" (Tufte 2006).

There are many reasons we might want to include visual displays in our writing; however, the two most common purposes in journal articles, internal memos, and reports are (1) to easily communicate the experimental setup or (2) to easily communicate the results of the experiment. Although there are many visual display tools that can be used to accomplish these purposes, let's just look at a few of the more common and effective visual tools.

3.4.1 Experimental Sketch

We want to get in the habit of thinking about, considering, and distinguishing everything that might have an impact on the experiment that we performing. Accompanying any experimental report or paper, there is typically a sketch of the experiment. Occasionally, a photograph of the experimental setup is used. Photographs can be distracting and actually take away from what we want to communicate. A sketch or even a block diagram such as a process flow diagram, can show very specific views. Labeling can provide the necessary amount of detail. The remainder of the essential details can be reserved for the text. If there are many parts, a legend can be used. Compare Figures 3.1 and 3.2 of a belt furnace used to braze and anneal metal parts. The simple sketch allows us to identify the essential elements in the furnace. The arrows can indicate the direction of motion. This is more difficult to show with the photograph than a simple sketch. The belt conveys the parts through the furnace at a set speed. The parts to go through a rapid thermal process are loaded

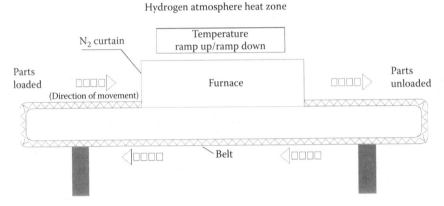

FIGURE 3.1
Sketch of a belt furnace.

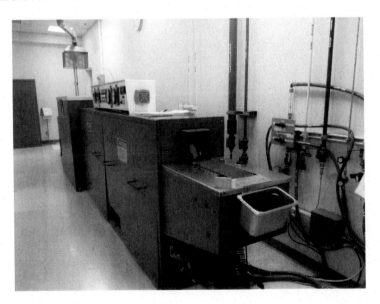

FIGURE 3.2
Photograph of a belt furnace manufactured by C. I. Hayes. (Courtesy of Coherent, Inc.)

on one end and unloaded on the other end after going through the hydrogen furnace. The parts are subject to a temperature profile determined by the belt speed and the set point temperature inside the furnace. The contrast in the amount of information communicated by these simple figures is striking. We can learn so much more from the sketch than the photograph.

3.4.2 Process Flow Charts

We can think of a process as anything that takes inputs to achieve some desired result(s) or output(s) (ISO 2008). In an experimental setting, an experiment may include many processes. Before we begin to collect data, we want to understand the overall process as well as the smaller processes embedded within the larger process. Deciding when, where, and how to collect data can be as important as the data collection itself. A process flow diagram is a visual representation of all the major steps and decision points in a process. This graphical tool can help us understand the process behavior better, identify critical or problem areas, and areas where important improvements might be made. Although the process diagram may never be viewed by anyone except us, the experimental designer, and our colleagues, it is a helpful tool in identifying where and why certain decisions are made. The diagram can show the flow of samples which may take different preparations or measurements.

Almost all processes can be represented by a process flow diagram. The level of detail can be adjusted based on the skill level of the persons using the chart or based on the level of abstraction we want to create. We could think of the process flow diagram as our storyboard. Figure 3.3 provides a set of symbols that may be valuable in communicating processes steps. Whether we use the symbols in Figure 3.3 or create our own, this tool allows us to create for ourselves a crisp, clear sequence of events for our experiments. Again, we may never publish this diagram or show it to anyone else; however, it can be a powerful analytical tool to link the experimental actions and events in our experiments. If an experiment doesn't follow the same flow as other experimental runs, maybe that can account for an anomaly or discrepancy that shows up later or maybe it indicates a level of robustness in the absence of anomalous results.

A process flow diagram can be thought of as a flowchart for a process. We can create flowcharts (process flow diagrams) for everyday common activities like cutting the grass or operating the washing machine, and for more complex activities we might perform in our experiments like operating a belt furnace or a scanning electron microscope. The level of

	Start Stop	Report trouble Report machine inoperation
	Decision point	Complete/incomplete Accept/reject Ready/not ready
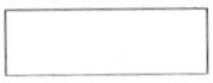	Activity or action	Fill breaker Turn on furnace
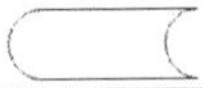	Record data	Write observations in a lab notebook or computer
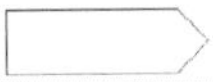	Delay	Heat Cool
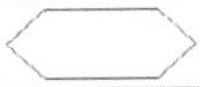	Preparation	Prepare a sample Get ready for next steps
	Direction of flow for process	
	Connector to another page or part of the process	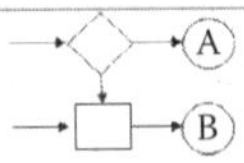

FIGURE 3.3
Common symbols used in process flow charts.

detail required depends on several factors, such as the intended users of the flow diagram and the criticality that the steps be performed exactly as specified. For example, if we are creating a flowchart for a chemistry experiment, the order in which chemicals are added may be important. Creating a process flow diagram can provide step-by-step instructions for the experiment. Figure 3.4 is an example from a chemistry experiment.

3.4.3 Input–Process–Output Diagram

> Good science involves understanding all the factors that might enter into a measurement.

> **Lisa Randall (2011)**

A process is defined as the merging of inputs to create an output. The output is a measure of performance with respect to the expected outcome (customer requirements is an example of an expectation). Inputs are everything, yes, absolutely everything, that might have an effect on the process. The Input–Process–Output diagram is a great tool for demonstrating a thorough understanding of the process. Figure 3.5 is an example of a generic template for an Input–Process–Output diagram.

In industry, performance measures are typically things like cost, time, defects, error rate, throughput, cycle time, cost of poor quality, on time delivery, etc. Additionally, in a classroom as in industry, the performance measure or expected outcomes might be what we are attempting

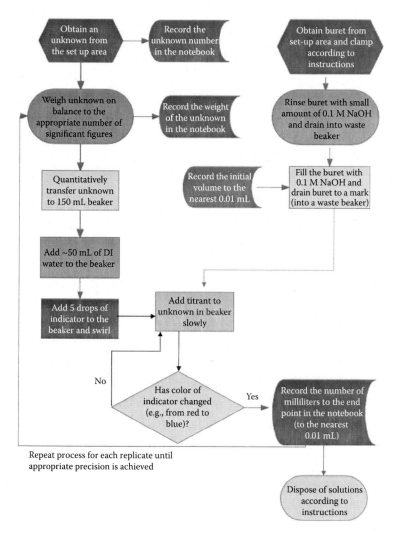

FIGURE 3.4

Example of a chemistry experiment flow chart. (Courtesy of Professor D. Nivens, 2016, http:/www.chemistry.armstrong.edu/nivens/Chem2300/flowchart.pdf.)

to accomplish or show in our lab—corrosion, joining, sealing, etching or deposition, a chemical reaction, etc. It is important that we know what we want to accomplish. We need to know the desired outcome. These outputs must be metrics. In other words, these outputs must be measureable. If we can't measure it in some way, how do we know if the process is working? We don't know what we don't know … So let's measure it!

Let's take another example from materials science and engineering, which involves establishing interrelationships between the structure and

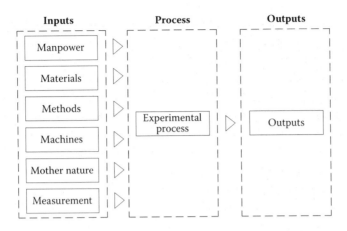

FIGURE 3.5
Generic Input–Process–Output diagram.

properties of materials. The structure and properties of materials determine the performance of the material. The structure determines the properties of the materials. (For example, face-centered cubic materials are more ductile in general than body-centered cubic. Fine grained materials are stronger than course grained materials.) The properties can be altered by changing the composition and/or by processing. The structure and properties (determined by *composition and processing*) will determine the final performance of the material. The performance of the material will depend on the type of mechanical loads it's subjected to, the type of environment it's exposed to, and the different combinations of temperature, stresses, and environmental effects the material experiences. Testing and experimentation are used to determine the structure and properties of materials (Callister and Rethwisch 2008). Figure 3.6 is an example of Input–Process–Output diagram for a braze process. A braze process is used to join two materials by melting a filler metal into the joint. Brazing is similar to welding and soldering. Welding requires local melting of the materials to be joined. Soldering can be performed at lower temperatures and allows a greater distance between the parts to be joined. Brazing creates a very strong join between either the same metals or different metals.

3.4.4 Infographics

An ill-specified or preposterous model or a puny data set cannot be rescued by a graphic or calculation, no matter how clever or fancy.

Tufte (2001)

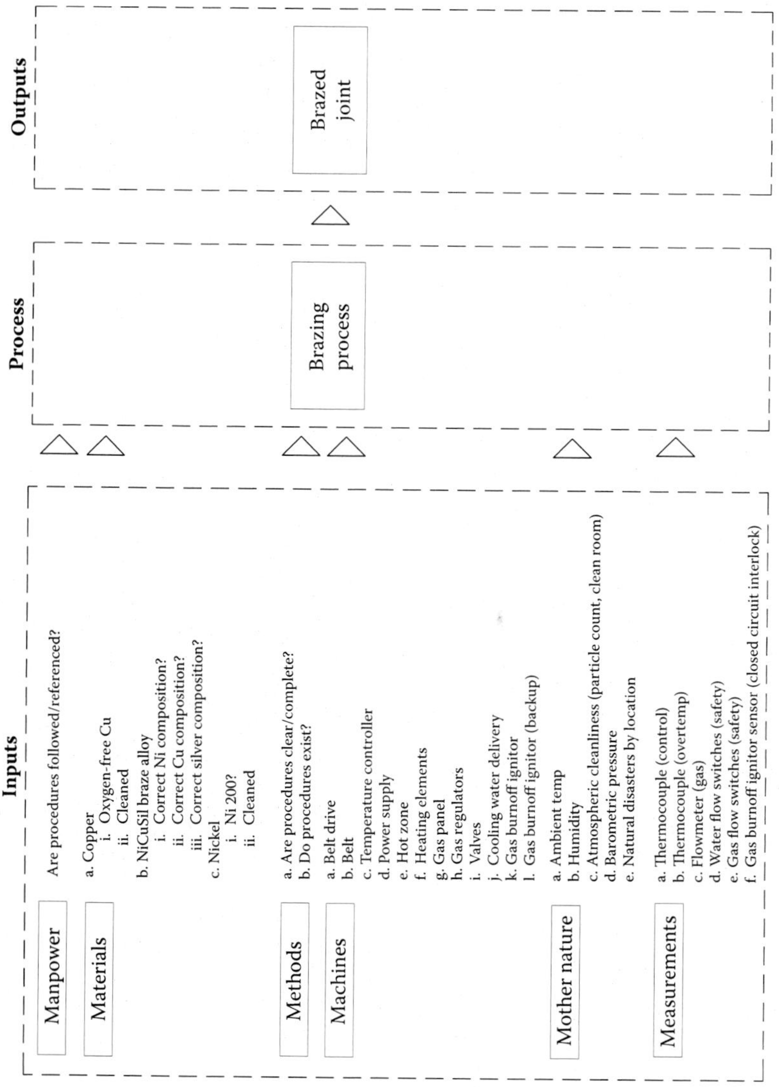

FIGURE 3.6

Example of an Input–Process–Output diagram for a braze process.

Infographics are relatively new data visualization tools. An infographic is a graphic that provides an explanation and/or information. Typically, an infographic is a combination of text and graphics. Most rail systems today use an infographic to display each of the stations. We can use infographics to explain complex processes. The infographic is a good all-in-one tool for communicating basics concepts. Figure 3.7 shows the basic processes that occur during chemical vapor deposition. We now see seven of the basic processes illustrated in the graphic, we know roughly where they occur inside the processing chamber, and we get the basic ideas of the terminology used by deposition scientists and engineers. For those of us familiar with chemical vapor deposition, we know that the process is much more complex. Not all processes are captured in this graphic. However, by introducing this graphic first to explain chemical vapor deposition, we can now discuss other processes that can occur during the process. From this graphic, we could launch the idea of a *mean free path* and point to possible alternative processes that our species may take as a result of their *mean free path* of different species. For example, we can explain that this chamber is in vacuum, which means that we have removed the majority of the particles in the chamber. The *mean free path* is the distance a species can travel before it interacts with another particle inside the chamber.

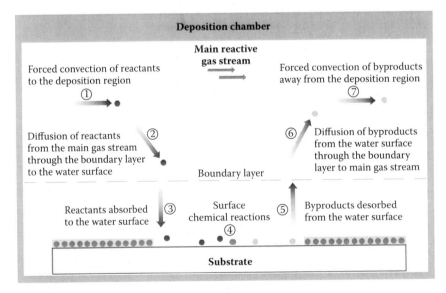

FIGURE 3.7
Infographic showing the processes involved during chemical vapor deposition. (From Plummer, J.D., Deal, M.D., Griffin, P.B., *Silicon VLSI Technology: Fundamentals, Practices and Modeling*, Prentice Hall, Upper Saddle River, NJ, 2000.)

The lower the pressure inside the chamber, the longer the *mean free path*, which means that the electrons, ions, or molecules can travel further inside the chamber before reacting. The *mean free path* will affect each of the seven processes illustrated in the graphic.

3.5   COMMUNICATING EXPERIMENTAL RESULTS

Visualization provides insight that cannot be appreciated by any other approach to learning from data.

William Cleveland (1994)

3.5.1   Components of Graphs

While lying in bed one morning in 1636, the mathematician and philosopher Rene Descartes watched a fly crawling on the wall. As he watched, it dawned on him that the path the fly was making on the wall could be captured numerically. He noticed that the fly was initially 10 inches above the floor and 8 inches from the left edge of the wall. A moment later the fly was at 11 inches above the floor and 9 inches from the left edge. Descartes drew two lines at right angles, a horizontal line to represent the floor and a vertical line to represent the left edge of the wall, which is equivalent to the length of floor to ceiling. The two lines intersected where the walls met. As long as the fly was walking on the wall, its path could be traced precisely—a certain number of inches from the floor and a certain number of inches from the left wall. Descartes translated the idea of latitude and longitude for identifying a location on earth relative to the poles. The notion of latitude and longitude to identify a global position had been around since the 3rd century BC. However, Descartes' realization was that two quantities could be used to represent a relative position. Descartes was able to construct a grid relative to the two lines. The idea of the graph as a "sophisticated abstraction" created a "conceptual revolution." Descartes "showed that algebra and geometry were two languages that described a shared reality" (Dolnick 2011). The coordinate plane (grid) that is created with the two lines Descartes envisioned was named in his honor, the Cartesian coordinate system. Each location on the grid is defined with two identifiers, which provide the location of the data point relative to the two lines known as the x axis and the y axis (Johnson and Moncrief 2002).

Just as Descartes was able to map out the fly's path with a graph, we can use graphs to reveal our data at multiple levels of detail. Graphs are a wonderful way of understanding and sharing information. University of Colorado Physics Professor John Taylor wrote "… drawing graphs helps you understand the experiment and the physical laws involved" (Taylor 1982).

Every professor or manager who asks us to create a graphic will have his or her own guidelines for what is important and essential to include in a graph. However, there are four basic elements to any graphic: labeling, scaling, the data itself, and possibly the trend line. Before proceeding with these essential elements, let me make a quick comment about gridlines, shading, 3D bars, or other effects: just don't. The purpose of a graphic is to communicate information; be as spartan as possible with all the extra stuff. Three-dimensional effects tend to distort the data, misleading the reader (Klass 2012, Tufte 2006). The data, not the methodology of plotting data, are the crux of the graph. Let's look at each one of these elements.

From my experience teaching and managing young engineers and scientists, improper and/or incomplete labeling tends to be the most overlooked or ignored part of a graph. Proper labeling is critical in communicating exactly what we are plotting. Although there will be specific guidelines about what to include and what not to include in our specific situation. My advice is to overlabel just to be on the safe side. It is better to label too much versus not enough; however, balance this with having the labeling be a distraction from the communication.

Labeling includes the title, the axis labels, the legend, and, when the graphic is incorporated into a report or paper, the figure caption (see Figure 3.8). In journal articles and/or lab reports, the title can be forsaken for the figure caption to avoid redundancy and use the real estate on the

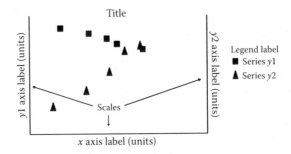

FIGURE 3.8
Example of all the labels that may need to be included in the graph. The legend is crucial when plotting more than one variable.

paper completely for the graph. In certain situations in presentations, this may also be the case. If our slide contains only the one graphic, the title may be included at the top of the slide rather than duplicated both at the top on the slide and above the graph. However, the title or the figure caption, whichever we use, should define precisely the data being plotted. It should provide the reader/viewer with a very brief, clear message about our data and/or the statement we are making about the data. In rare cases, we may want to have both a chart title and a figure caption, but this should be the exception not the rule.

A pet peeve of engineering managers and professors is axes without labels and/or units of measurement. The graph should clearly convey the information contained in the graph; this isn't possible if the reader cannot tell at a glance exactly what is plotted. The standard convention in most scientific journals in the United States is to label each axis and include the units in parentheses immediately following the variable name. This doesn't mean that we cannot be creative with axis placement and labels to have the graph more effectively reach the intended audience (Knaflic 2015).

If we are showing more than one data set on a graph, a legend is necessary. As we choose the symbols for our two sets of data, keep in mind that many times, our graph or chart may be reviewed on paper (for those of us in the older generation) and the printing on that paper may be in black and white. Use symbols that are large enough to be distinguished from one another, so they can be easily read if that beautiful color graph is printed on a dot matrix printer in black ink.

There are times when we may want to label some of the actual data points. For a large number of data points, this practice can get messy and confusing. However, if we have a point to make about a data point, then labeling that data point or a few points can be an effective way to illustrate comparisons. Figure 3.9 gives an example of the ineffective and effective use of labeling data points.

When choosing a scale for our x and y axes, we want to fill the majority of the space in the chart. This applies to both axes. Squeezing all the data into 20% of the y axis isn't good. Center the chart on the data. Adjust the axes to accomplish this goal. There are times when it doesn't make sense to have the origin be (0,0), but if it doesn't support communication of the message of the graph, it isn't necessary. Notice that in the graph in Figure 3.10a, we have started the graph at the origin, although there are no temperature data points below 400 K. Also, notice that the mass flow rate for the fuel consumption is so small that we can see very little about

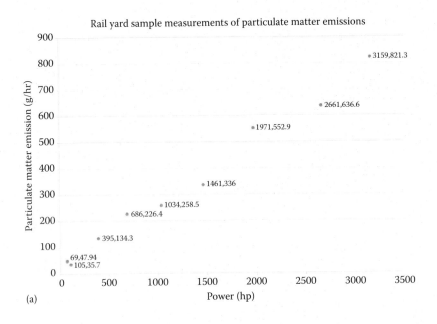

(a)

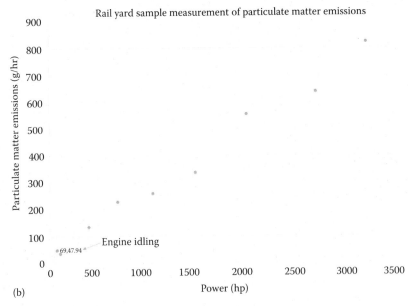

(b)

FIGURE 3.9

(a) Example graph where all data points are labeled. Without additional information, this is ineffective and distracting. (b) Example of graph where one data point is labeled that provides specific details of particulate matter emissions during the engine idle condition. (From Filippone, C., *Diesel—Electric Locomotive Energy Recovery and Conversion: Final Report for Transit IDEA Project 67*, 2014, http://onlinepubs.trb.org/onlinepubs/IDEA /FinalReports/Transit/Transit67.pdf.)

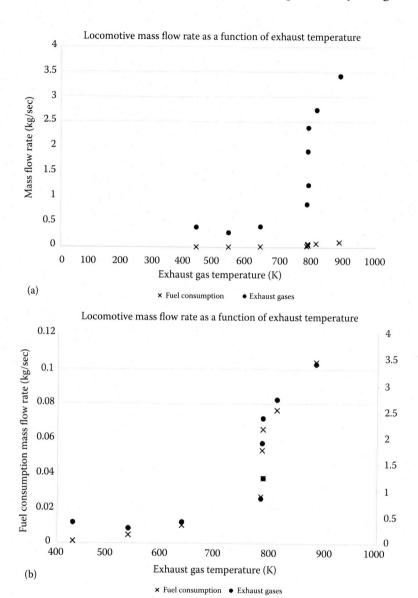

FIGURE 3.10

(a) Graph of mass flow rate from locomotive fuel consumption and exhaust on the same graph and scale as a function of locomotive exhaust gas temperature. (b) Graph of mass flow rate from locomotive fuel consumption and exhaust on the same graph but different scales as a function of locomotive exhaust gas temperature. (From Filippone, C., *Diesel—Electric Locomotive Energy Recovery and Conversion: Final Report for Transit IDEA Project 67*, 2014, http://onlinepubs.trb.org/onlinepubs/IDEA/FinalReports/Transit /Transit67.pdf.)

what's really going on. This same data is replotted in Figure 3.10b, and we can see that the mass flow rate as a function of temperature is the same for both fuel consumption and exhaust. We see a steep increase around 800 K. There are other cases where we may want to show two different dependent variables on the same graph. Scaling is important in these cases to effectively communicate the data. Figure 3.11a and b demonstrates the same concepts with two different variables. The scale we use can highlight the information we wish to communicate or obscure vital information.

How we go about plotting our data should be determined by the data. The majority of scientific data will be plotted in scatter plots. Scatter plots easily show the actual data points and allow for comparison of more than one set of data. Scatter plots are the most common types of graphics used in journals, followed by contour plots (The contour plot images a 3-dimensional surface by plotting constant z slices, called contours, on a 2-dimensional format. The (x,y) coordinates are connected where that z value occurs.). Scatter plots are nice because we can actually see the effects of the two variables plotted; while contour plots, as well as 3D plots, give us a "feel" for the data but tend to be less specific.

3.5.2 Introduction and Examples of Useful Graphical Tools

How and when to use each type of graphic may be confusing. Again, what do we want to show? The majority of graphics are used to display (1) trends, (2) distributions, (3) compositions, (4) processes, and/or (5) locations.

Table 3.2 compares several of the more commonly used scientific graphics to illustrate the differences. Each of these broad categories listed contains multiple subcategories of chart types. The advantage of tables and graphics is the ease with which they allow comparison, correlations, or patterns to become obvious (Khorasani 2016). "A table of numbers might contain the identical information, but a table muffles the patterns and trends that leap from a graph" (Dolnick 2011).

3.5.2.1 Pie Charts

Pie charts are rarely used in hard science publications. Even business leaders are recommending that they not be used (Knaflic 2015). I don't see much use for these in engineering or science. The data in a complex pie chart can typically be expressed with a different type of graph while a simple pie chart can be expressed in a table. There is nothing wrong with using pie charts,

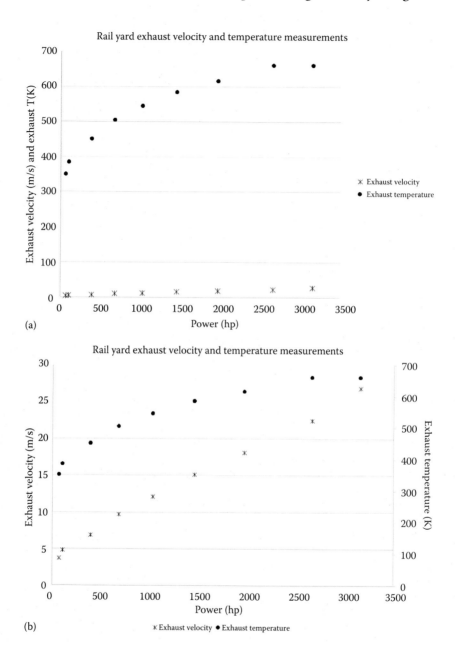

FIGURE 3.11

(a) Graph showing exhaust temperature and velocity measurements on the same graph and scale as a function of engine power. (b) Graph showing exhaust temperature and velocity measurements on the same graph and scale as a function of engine power. (From Filippone, C., *Diesel—Electric Locomotive Energy Recovery and Conversion: Final Report for Transit IDEA Project 67*, 2014, http://onlinepubs.trb.org/onlinepubs/IDEA/FinalReports/Transit/Transit67.pdf.)

TABLE 3.2

Comparison of Various Graphical Techniques and When Their Use Might Be Appropriate

Graphic	Goal	Comments	Examples
Pie chart	Composition	Emphasize relationship to whole	Don't use, but if you must, express a qualitative relationship
Scatter plots	Trends, comparisons	Emphasize relationships	Trends in space or time, relationships
Photographs/ micrographs	Distributions, trends, composition	Only include object of interest	Process results, e.g., etch depth or profile, cross sections
Schematic or process flow diagram	Processes, locations	Simplify complex processes	Complex experimental setup or process flow
2D or 3D contours	Trends, locations, compositions, distributions	Contrast two results	Surface roughness, topographical maps
Tables	Comparisons	Compare categories of results	Comparison between two experiments
Histogram or bar chart	Distribution, composition, trends	Emphasize categories/ groups in data	Summarize a large data set

should we feel that is the best way to express our data. Pie charts might be useful where we have a few (typically less than six) categories of data and the relative size of these categories as a part of the whole is important. With similar percentages in the categories, a bar chart might be a better tool for effectively communicating the results. I went through almost a complete year of issues of the journal *Applied Physics Letters*, counting and categorizing all the figures for the year. I didn't find one example of pie chart in the data. My findings are summarized in Figure 3.12.

Distributions or trends can be seen easily with histograms or frequency diagrams and scatter plots rather than pie charts. Scatter plots are great for displaying data collected over time or over a distance, allowing the reader to visualize the distribution. Before selecting a pie chart to communicate data, we should consider other more effective charts (Knaflic 2015).

3.5.2.2 Histogram

A histogram depicts frequencies of numeric data whose purpose is to provide a pictorial summary of a data set. In other words, the histogram

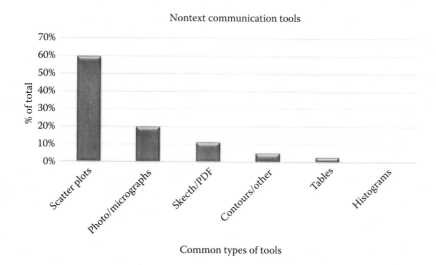

FIGURE 3.12
Bar chart showing the most common types of figures in *Applied Physics Letters* in 2000.

provides us with a frequency distribution, which gives us a picture of the data. Histograms can be used to present both numeric and nonnumeric data occurrences.

A histogram is a beautiful graphic. It is made up of frequency on the *y* axis and intervals on the *x* axis. A display in this form provides a synopsis or summary of the data. We see how the values are distributed. By looking at the histogram, we can identify roughly the center and extremes of our data. We can see how the data set varies and any symmetry or lack thereof about the center point. The height of each bar tells us how many measurements were observed to be in the interval of that bar. The exact appearance of any histogram depends on the chosen class for the horizontal axis. Spreadsheet and graphing programs will use a default algorithm for setting up a histogram, but ultimately, we are responsible for adjusting the number of intervals and range to allow for the necessary interpretation and analysis of our data.

If our data is numeric, one way to display it using a histogram is by counting the number of times each value occurs. The number of times a value occurs is called the frequency. The area of each column in the histogram is proportional to the frequency of the values within that cell. As you might imagine, if we continue to collect more and more data, the height of each of these cells would begin to form a smooth curve. This curve is called a frequency distribution curve. We can save the details for a later chapter.

TABLE 3.3

List of 30 Hardness Measurements of 30 Different 304 Stainless Steel Discs

89.2	85.6	85.5	83.6	84.5	86.4
85.3	83.2	87.9	85.1	85.1	87.5
85.3	85.5	86.4	83.8	84.5	87.2
86.7	85.5	87.3	80.7	82.1	86.2
83.8	84.1	88.6	82.7	86.2	87.5

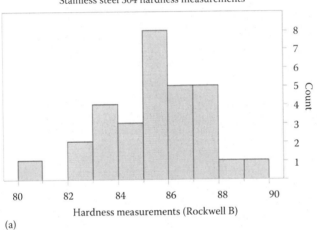

(a)

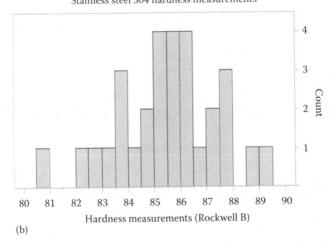

(b)

FIGURE 3.13

Histogram of the hardness measurements on 304 stainless steel samples with bin sizes of (a) 2 and (b) 1.

Let's look at a data set of 30 numbers, listed in Table 3.3 and displayed in Figure 3.13. Table 3.3 has 30 measurements of the hardness of 30 different 304 stainless steel discs using a Rockwell B Hardness Tester. What can we see just from looking at the data? From the histogram, we can see if the data vary wildly. We might also look through the data set and pick out the minimum value and maximum values. In spite of this, we don't really get a feel for the data or what might be going on with the experiment. Depending on how large the data set is, neither of these values really gives us a feel for the data set. A histogram gives a better synopsis of the data. We see how the values are distributed and we see the center of the data set (roughly anyway). We see how the data vary and any symmetry or lack thereof about the center.

3.5.2.3 X–Y Scatter Plots

Histograms deal with a single variable. Many times, we have more than one variable. This is bivariate data meaning two variables. These data come in ordered pairs, for example (x, y). The purpose of a scatter plot is to visually study the relationship between two variables. Be careful about drawing cause and effect relationships based strictly on how a graph looks (recall the number of patents granted versus life expectancy in the US example).

Labeling of the axes is somewhat arbitrary, even for us as engineers and scientists. If we know that one variable is the dependent variable or response variable, this is graphed on the y axis. The independent variable is graphed on the x axis. The axis should always have the name of the variable and the units of measurement in parentheses. Scaling of axes can be somewhat arbitrary as well. Remember that the goal of the graph is to allow the viewer to see a relationship between the variables if one exists. Be careful connecting points with line segments in a scatter plot. We should be convinced of the connection between the two points before doing this. Similarly with trends, a trend line may have a slope that indicates change or remain flat. Figure 3.14 shows a scatter plot of calendar year on the x axis and percentage of patent applications granted on the y axis. Notice a trend line has been drawn through the data points. We see a very slight downward trend to the line showing that roughly 59% of patent applications are granted each year. This has been a consistent trend with only a slight decrease over the years.

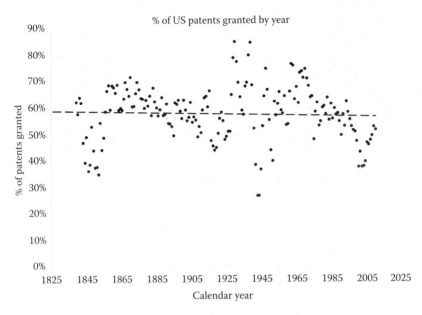

FIGURE 3.14

Scatter plot of the % of patents granted by years in the United States. Roughly 58% of the patent applications received are issued patents. (From United States Trademark and Patent Office, U.S. Patent Activity Calendar Years 1790 to the Present: Table of Annual U.S. Patent Activity Since 1790, https://www.uspto.gov/web/offices/ac/ido/oeip/taf/h _counts.htm, 2016.)

3.5.2.4 Time Series Data

Scatter plots can be used to represent data collected over time. The purpose of this chart is to monitor a system or process and to detect any meaningful changes in the process over time. This type of chart has a number of different names, one being a run or process chart. How do we detect a change? There are no hard and fast rules; however, it is generally expected that we'll have roughly the same number of points above and below the mean (average). If we see seven in a row on one side of the average, something might be happening. Be careful in reading a run chart. If we chase every slight variation in our data, we might actually miss significant changes.

A time series graph is like a scorecard to show whether we are changing or staying steady. The most effective way to use this graph is in conjunction with other graphs like a histogram. Figure 3.15 shows a dramatic increase in patent applications in recent history. This type of behavior is often called a hockey stick graph.

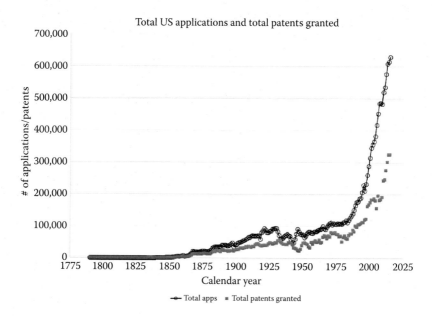

FIGURE 3.15

Time series data showing the dramatic increase in patent applications and patents granted in recent history. (From United States Trademark and Patent Office, U.S. Patent Activity Calendar Years 1790 to the Present: Table of Annual U.S. Patent Activity Since 1790, https://www.uspto.gov/web/offices/ac/ido/oeip/taf/h_counts.htm, 2016.)

3.5.2.5 Tables: When and Why

Tables are the most popular form of communicating data, often in the form of spreadsheets accompanied by a few graphs or charts. The use of tables seems to confound many new engineers and scientists. There are two common rookie mistakes. Both listing measurement results in a sentence or series of statements and bulking up a report with pages and pages of data tables are tedious and annoying for our readers/audience. No matter what level of class I teach to undergraduate seniors or beginning graduate students, or level of engineer I hire (BS, MS, or PhD), many times, the first report I read will have page after page of embedded data in tables in the body of a lab report with corresponding graphs to go along with it. Don't do this to a manager or instructor. Tables displaying every data point that we collect do not belong in the body of any report or paper (unless there is a specific request that we do this). Small data tables can be included if there is a point to make in the report. The more appropriate place to present large data sets would be a table in an appendix.

Tables are great tools for summarizing, comparing, and presenting data in order to communicate results. See Tables 3.1 and 3.2 summarize information. Tables are an effective tool for presenting different experimental conditions, results, and/or why the testing was performed. Use tables to extrapolate the information contained in the text. There are times when it is easier to present a summary of some data set in a table rather than a graph. Compare Tables 3.3 and 3.4 in this chapter. Both tables contain exactly the same data. Table 3.4 gives us information about the samples. If we knew that the sample groupings were from six different material lots (A, B, C, D, E, and F), would this make the information more valuable to consume real estate in the main body of our report?

Much of the same rationale behind graphic displays also applies to tables. All columns and rows should be clearly labeled and easy to read. The labels should provide information about exactly what is contained in the table and any associated units. Tables should have captions in the text, just as with figures. Be aware and conscious of the number of significant digits used in tables. Just because Excel or the calculator will give 10 digits when two numbers are divided, doesn't mean the data are accurate to 10 digits, unless, of course, it is. I can only think of a few situations where this degree of accuracy would even be important to present. For example, if we worked for NASA or a national lab, there might be occasions where writing gravity out to 10 or more decimal places would make a difference. I stress this here and in the next chapter because it happens far too often. We get caught up the in the calculations and the excitement of our findings and the detail of the digital accuracy are glossed over.

TABLE 3.4

30 Hardness Measurements of 30 Different 304 Stainless Steel Discs Emphasizing Groupings

Sample	Hardness (Rockwell B)	Sample	Hardness (Rockwell B)	Sample	Hardness (Rockwell B)
A1	89.2	B1	85.6	C1	85.5
A2	85.3	B2	83.2	C2	87.9
A3	85.3	B3	85.5	C3	86.4
A4	86.7	B4	85.5	C4	87.3
A5	83.8	B5	84.1	C5	88.6
D1	83.6	E1	84.5	F1	86.4
D2	85.1	E2	85.1	F2	87.5
D3	83.8	E3	84.5	F3	87.2
D4	80.7	E4	82.1	F4	86.2
D5	82.7	E5	86.2	F5	87.5

3.6 IMPORTANCE OF CONCLUSIONS

University of Colorado Physics Professor John Taylor wrote, "Performing an experiment without drawing some sort of conclusion has little merit" (Taylor 1982). In my experience both in industry and academia, I continue to see many reports that omit any discussion of the findings and/ or fail to draw conclusions, especially from new scientists and engineers. Drawing conclusions seems to work one of two ways: either there are no conclusions and the presenters stop midsentence or the conclusions are so all-encompassing that the presenters throw out all they know of statistically significant results. As scientists and engineers, especially in the laboratory or early career, when data become available, it "must be interpreted through the construction of a theory that can explain" the results (Weisberg 1993). The conclusions and/or discussion section is an integral part of work. This section brings the experiment into perspective for all readers. In reality, the conclusion/discussion section is the most important section of any write-up/presentation, not just an afterthought.

There are a few basic guidelines for writing or presenting a good experimental conclusion.

- Begin and end on a positive note. Even if the experiment wasn't successful or completely successful, we can still highlight what we did learn from doing the work.
- Compare results to literature or tribal knowledge. We should relate our work to what others have done or what is thought.
- Likewise, compare results to initial hypothesis or problem statement. Everyone reading will want to know if the results matched our initial hypothesis or answered the question.
- Quantify/qualify results by highlighting sources of error. We should never hide from this. Openly identifying errors will allow others to have confidence in our work.
- Describe additional experiments that would improve results.

The primary purpose of the conclusion/discussion section(s) is to describe how our findings relate back to and/or solve that basic problem. The discussion section should be supported by and relate the initial problem directly back to our experimental findings. The discussion/conclusions should explain what the results mean. There are times when it is important to extrapolate or generalize from the observations we've made. In these cases, we should make it clear that is exactly what we are doing. If the conclusions are a confirmation of prior work, we should reference the others' work.

3.7 KEY TAKEAWAYS

The objective of communication in a lab or work setting is to convey ideas about our work, actions we have taken, and conclusions we have drawn. We can achieve this objective with text, tables, graphic displays, or a combination of all three. Therefore, the objective of any text, table, or graphic is to communicate. We want to create as simple and clear of a message as we can with our data. The less our audience has to struggle with to understand our work, the more confidence they can have in our abilities as experimenters. A chart (or graph) that is confusing or in any way unclear will not be effective in getting our message across and could potentially erode any confidence in us and/or our experiment. We have a choice about how to communicate our work, and we must decide in each presentation setting which method(s) will be most effective and engaging. We want our work not only to inspire interest but confidence in our experimental abilities.

P.S. Take some time to watch scientists and engineers giving TED talks (Technology, Education, and Design) talks. I personally find that no matter the topic of their work, I am engaged and interested. Our technical presentations may need to be more detail oriented but it doesn't mean we can't learn from this style of presentation.

REFERENCES

Anderson, C. 2016. *TED Talks: The Official TED Guide to Public Speaking.* New York: Houghton Mifflin Harcourt.

Callister, W. D. and D. G. Rethwisch. 2008. *Fundamentals of Materials Science and Engineering: An Integrated Approach.* 3rd Ed. New York: John Wiley & Sons.

Cleveland, W. S. 1994. *The Elements of Graphing Data.* Summit, NJ: Hobart Press.

Deming, W. E. 1982. *Out of the Crisis.* Cambridge, MA: Massachusetts Institute of Technology, Center for Advanced Engineering Study.

Dolnick, E. 2011. *Clockwork Universe: Isaac Newton, the Royal Society and the Birth of the Modern World.* New York: HarperCollins.

Gallo, C. 2014. *Talk Like TED: The 9 Public Speaking Secrets of the World's Top Minds.* New York: St. Martin's Press.

Hasson, U., A. A. Ghazanfar, B. Galantucci, S. Garrod, and C. Keysers. 2012. Brain-to-Brain Coupling: A Mechanism for Creating and Sharing a Social World. *Trends in Cognitive Sciences* 16(2):114–121.

Hoffman, R. 2014. The Tensions of Scientific Storytelling: Science Depends on Compelling Narratives. *American Scientist* 102(4):250.

Houston, S. D. 2008. *The First Writing: Script Invention as History and Process.* Cambridge: Cambridge University Press.

Hsu, J. 2008. The Secrets of Storytelling: Why We Love a Good Yarn. *Scientific American Mind* September 18.

ISO (International Standards Organization). 2008. Document ISO 9001:2008. *Quality Management Systems—Requirements.* http://www.iso.org.

Johnson, A. V. and J. W. Moncrief. 2002. Descartes and His Coordinate System. *Mathematics.* http://www.encyclopedia.com.

Khorasani, F. 2016. Private communication.

Kim, Y. J. 2015. How Undergraduate Journals Foster Scientific Communication. *Public Library of Science.* http://www.plos.org.

Klass, G. 2012. *Just Plain Data Analysis: Finding, Presenting and Interpreting Social Science Data.* New York: Rowman & Littlefield.

Knaflic, C. N. 2015. *Storytelling with Data: A Data Visualization Guide for Business Professionals.* New York: John Wiley & Sons.

Livio, M. 2013. *Brilliant Blunders: From Darwin to Einstein—Colossal Mistakes by Great Scientists That Changed Our Understanding of Life and the Universe.* New York: Simon & Schuster.

Mlodinow, L. 2008. *The Drunkard's Walk: How Randomness Rules Our Lives.* New York: Pantheon Books.

Newton, I. 1999. *The Principia: The Mathematical Principles of Natural Philosophy.* Translated by I. Bernard Cohen and Anne Whitman. Berkeley, CA: University of California Press.

Randall, L. 2011. *Knocking on Heaven's Door: How Physics and Scientific Thinking Illuminate the Universe and the Modern World.* New York: HarperCollins.

Sobel, D. 1999. *Galileo's Daughter: A Historical Memoir of Science, Faith and Love.* New York: Bloomsbury USA.

Stanton, A. 2012. Talk presented at TED2012. http://www.ted.com/talks/andrew_stanton _the_clues_to_a_great_story.html.

Taylor, J. R. 1982. *An Introduction to Error Analysis: The Study of Uncertainties in Physical Measurements*, 2nd Ed. Sausalito, CA: University Science Books.

Thaler, R. H. 2015. *Misbehaving: The Making of Behavioral Economics.* New York: W. W. Norton.

Tufte, E. 2001. *The Visual Display of Quantitative Information.* Cheshire, CT: Graphics Press.

Tufte, E. 2006. *Beautiful Evidence.* Cheshire, CT: Graphics Press.

Tukey, J. W. 1977. *Exploratory Data Analysis.* Boston: Addison-Wesley.

Weisberg, R. W. 1993. *Creativity: Beyond the Myth of Genius.* New York: W. H. Freeman and Company.

Williams, H. 2013. Storytelling and Science: The Unifying Theory of 2 + 2. *The Berkeley Science Review.* May 27, 2013. http://berkeleysciencereview.com/.

4

Introducing Variation

If statistics is detective work, then the data are the clues.

Charles Wheelan

Making measurements and collecting data are not the goals of engineers and scientists. Making measurements and collecting data are merely a means to and end. The purpose of measurements and data collection is to help us (our lab partners or team, company, or manager) make informed decisions—for example, decisions about whether a product is shipped may depend on whether a process or tool is working or needs improvement. The job of an engineer is to make decisions or recommendations about decisions—not just to collect data. The confidence others have in our experimental or problem solving abilities is a direct result of the choices we make and the data we collect.

Understanding the measurements and the data we collect are critical first steps in experimentation. We need to be able to effectively communicate the data we collect, but in order to do this, we must have agreements and understandings about the quality, quantity, type, and confidence of that data. For this reason, we need to discuss data and measurements early in our conversation about experimentation and problem solving. In this chapter, we will start from the beginning with data basics as a refresher and as a means of establishing a common way of languaging our results as data. Once we establish a conventional way of talking about our data, we can then examine measurements more closely. We know that all measurements (ideally) will contain some part signal and some part uncertainty (noise). Our confidence in the data, and therefore in our experiment, is often a measure of the ratio of the effect (signal) to the uncertainty (noise) or what is commonly known as the "signal-to-noise ratio." The higher the "signal-to-uncertainty" ratio, the more confidence we have. Next we will discuss opportunities for strengthening our data and then identify three distinctly different types of variation that contribute to uncertainty. The concepts in this chapter are essential for understanding the level of confidence our audience (which may be the senior managers or executives at the company, customers or professors) should have in both the measurements we make and the data we collect.

4.1 DATA CHAOS

One of the biggest problems we face as we try to solve the big problems of the twenty-first century is not the lack of data but data chaos. Over the years, our governments, health care organizations, and industry have collected heaps of data. The data might be expressed in English or metric units. Not only are the data expressed differently, but also these mountains of data are in file cabinets, basement boxes, and computers. The data may be gathered from many different specialists, different labs with different standards, using different protocols. The data can be handwritten or digital. A very small portion of that digital data is kept in well-organized databases. Much of digital data appears in an unstructured format while much of the paper reports are handwritten, scanned, and low-resolution. In the case of medical data, let's not forget about all those archived paper or audio files.

Miguel Helft provides a great example in an article he wrote for *Fortune* magazine (Helft 2014). Medical cancer data aren't collected systematically, and there are no standards for reporting the data. For example, data for albumin, a protein marker routinely measured in cancer patients, can be expressed in over 30 different ways. Albumin is just one marker. The real problem is that oncologists collect thousands of data points about each patient: from different blood markers, biopsies, genetic tests, magnetic resonance images, x-rays, etc. With each care facility and lab reporting data in different formats using different forms and storing the data differently, it will take dedicated efforts to make sense of all this. Innovation, from our modern conveniences to life-saving medical treatments, would only be science fiction without the ability to measure and control critical data in our experiments and research.

What is meant by data? The simplest synonym is information. In the sciences, we need data to meet four critical requirements: objective, comparative, representative, and useful information (Wheelan 2013). Let's separate each of these words for further discussion.

First, data are objective. Even if there is disagreement around what is interpreted from the data, there should be agreement about data itself. We see this in our lives each day. Scientists collect data of average temperatures around the world. These scientists are using calibrated, state-of-the-art equipment. The data are not in question; the planet is getting warmer. The earth has gotten 0.8°C warmer over the last 100 years. We also measure the amount different atmospheric gases of CO_2, CH_4, N_2O, etc. The subjective (nonobjective) part of this discussion is what the data mean. This has led many to put these (and other) facts together to conclude that our activities on this planet are the source of the temperature increase. Other scientists, albeit a minority (~3%), seeing the same data, do not draw these same conclusions. Therefore, the inclusion of the word *objective* in the definition of data is needed.

Data must also be comparative. "Type" is not a property of the data itself. "Type" is important to understand because it tells us something about the characteristics of the data. There are four common types of data with which we tend to be concerned in experimentation: measurement, nominal, ordinal, and locational. Properties and examples can be found in Table 4.1. It is important to know what kind of data we are looking for and

TABLE 4.1

Comparison of Quantitative Data Types

Data Type	Data Examples	Questions Answered by These Data	Numerical Expression
Measurement	Length, height, weight, volume, wavelength, power, time, temperature	How long? How much?	Any real number and units typically: 1.54353 meters, 7.954 W, 35.9 Joules, 10.5 m/sec, 0°C
Nominal	Frequency of occurrence	How many?	Integers: 3,1001,21
Ordinal	Ranking, ordered	What order were the students' grades in the class?	Integers: 1,2,3,4
Locational	Location	Where were cancer deaths by county in Idaho?	Real numbers and direction: 19.59852°N–155.5186°E

why we are collecting the data. In engineering and science, we deal with all of these types of data, but measurement data are the most widely used. For the most part, measurement data are quantitative (measured, numeric data) as opposed to qualitative (attribute, characteristic data), which may be numeric as in case of ranking (ordinal) data.

Measurement data are the most preferable type of data. Measurement data are quantifiable, continuous data, e.g., length, height, weight, volume, wavelength, power, time, etc. We get more information from an actual measurement as opposed to summarized data (statistics). Nominal data are the classification or categorization of data. Nominal data are quantitative, countable, discrete, or occurrence data, e.g., inches of rainfall, # of defects, # of failures, # of choices, # of birth defects, etc. Ordinal data are ranked data. Ranking birth order of den mates or states by the amount of rainfall are examples of ordinal data.

The final type of data is locational data. Locational data are used to answer the question "where?" and is typically found in concentration charts or measles charts. Locational data might be considered nominal data with a locator. By the way, measles charts are the locational graphs used to visually show where something is happening. For example, if we wanted, we could use a measles chart to show solar or wind power generation overlaid on a map of the Germany or traffic accident rates overlaid on a map of the Washington, DC, metro area.

The sciences deal with all types of data. Measured data are more informative, descriptive, and precise than counted data are. Since continuous data contain more information, they are preferred over discrete or discontinuous data. There are times when we have a choice about the type of data we collect. For example, if we are measuring a set of parts to compare against a drawing, we could measure the actual dimensions of the parts. For simplicity's sake, let's say we have a bag of 500 stainless steel washers. Once we had measured the thickness of all the washers, we would have 500 individual measurements. We would have a set of continuous data. We would be able to make calculations that represent the data set. On the other hand, we could also group the parts into discrete bins based on the measurements. This would give us countable data. For example, we can label the parts as "smaller than the specification," "within the specification," or "larger than the specification." In this case, we'd have a lot less information about the individual parts. We would only know which particular bin they belonged to and nothing else. In other cases, we do not have a choice about whether the data we collect are continuous or discontinuous.

For example, on the farm growing up, one of our morning chores was to collect eggs. We harvested eggs from the dozen or so chickens each day. Each morning, the number of eggs one of us kids had to gather was discrete or discontinuous. However, when my father asked me to calculate the average egg yield each week, I might have gotten a significant fractional number.

Data must be representative. There are many occasions where it is impossible to collect all the data that are available. *Population* is the term we use for the set or collection of all possible objects or individuals of interest. We are interested in a population because we want to draw some conclusions about the characteristics of that population.

In order to learn something about a population, we might collect a subset of the population data. A sample is a subset of population. In order to accurately represent a population, we need a random sampling of that population. When we talk about a set of data, we could be talking about a population or a sample. In the case of a sample, we want the data in the sample to be representative of the whole population or at least some larger group of that population.

Statisticians use a specific and particular language to distinguish and clarify statistics to represent population and samples. More often than not, we do not know the *true value* that describes a population. *True values* are not known; we, as scientists and engineers, use averages over repeated experiments to establish the reference value that we lazily refer to as the *true value* (Gauch 2006). How do we draw conclusions about a data set if we don't have a true value to refer to? We depend on calculated statistics to summarize the sample data available. These statistics include sample mean, median, range, standard deviation, variance, root-mean-square deviation, standard error, etc.

We'd like to be able to use sample data to draw conclusions about population. Ideally, we'd like to use one or two numbers to describe or represent our whole set of data. The most common approach is to describe the middle of the data and the variability or dispersion in the data. There are a number of rigorous mathematical implications that result from working with a population versus a sample of the population. We can leave that to future bedtime reading or a statistics class. Throughout this book, we will assume that we are talking about statistics that represent a sample. Describing the sample is a simpler and more cost effective way to represent the population. In a later chapter, we'll delve further into representative data.

Finally, data should be useful. Inclusion of irrelevant data in either reporting or analysis can lead to confusion or overly complex models. For example, recording what we ate for dinner or whether we were wearing our lucky socks is not likely to be an important experimental parameter. Don't laugh, I have read experimental reports and manufacturing protocols where equally silly factors and superstitions have been mentioned.

4.2 DATA BASICS

Before we delve into more advanced topics related to data, there are several data topics I'd like to review: significant digits and scales and units. I can't tell you how many college-level lab reports and even new engineers at all levels of education that I've reviewed with 10 to 12 significant digits in tables or scales and units have been omitted from reports. These may seem like elementary topics, but these are easy mistakes to make. The spreadsheet/workbook software applications make it easy to do calculations that default to displaying as many digits as there is room in the column. We focus on the numerical value and the calculation or measurements and forget the physical quantity that is of concern. At every step in our experimentation, we must simultaneously keep in mind both the big picture problem we want to solve and the ensuing details of calculations and/or measurements.

4.2.1 Significant Digits

Significant figures are defined as all non place holder digits in a number. What does this mean? It is probably easier to demonstrate with a few examples. Given the number 123.456, there are six significant figures in this expression. However, if the number were written as 123.4560, there may be either six or seven significant figures, depending on the whether the 0 in the ten-thousandths place was measured or is just a place holder. The same is true for numbers on the left of the decimal position; the number 10 could have one or two significant figures depending on whether the 0 was measured or whether it is just a place holder. Likewise, with 100, 1,000, and 10,000, the number of significant figures in each case could just be one or two in each case. In order to make the number of significant figures obvious, an alternative expression might be $10,000 = 1 \times 10^5$ for only one significant figure or $10,000 = 1.0 \times 10^5$ for two significant figures. Scientific notation eliminates any ambiguity in the number of significant digits.

Just as there is uncertainty in measured values, there is also uncertainty that carries over into calculated values. If we are performing some basic math function(s) on a number, we should report the least of the significant digits. For example, if adding the numbers 10.010 cm (four significant digits) and 10.5 cm (three significant digits), we would report only three significant figures. The resulting answer would be 20.5 cm. For addition and subtraction, the resulting answer should be rounded off to the last decimal place reported for the least precise number. Similarly, for multiplication or division, the number of significant figures in the product or quotient is determined by the expression with the least number of significant figures in the original numbers (Deardorf 2016). The uncertainty in calculated values should be carried over into the results, just as with the measured results, or at a minimum expressed with the correct number of significant digits.

The expressed or reported value for uncertainty will typically have one, or possibly two, significant digits. An experimentally measured value should be rounded to the number of significant digits that will make it consistent with the estimated uncertainty. For example, if we want to measure the mass of a 1991 US penny, it would be wrong to report the mass as $m = 8.93 \pm 0.4753$ g. We cannot know the uncertainty that accurately. To be consistent with the large unknown in the uncertainty, the uncertainty should be stated to only one significant digit. Following the rules of addition with significant digits, the mass should be expressed as $m = 8.9 \pm 0.5$ g.

4.2.2 Measurement Scales and Units

Have you ever missed a question on a test because you forgot to include the units? The difference between 2.45 as a number and 2.45 g, 2.45 km, or 2.45°C is critical. It is essential that units not be ignored. The units tell us exactly what the number physically represents—mass, distance, or temperature. The number has no meaning without the context provided by the unit. Achieving a solution of 2.45 may solve a math problem, but in a physics class, math is only a tool used to study physical properties.

Another important point related to the measurement scale is that scales are both arbitrary and relative. Measurement scales are relative to some established reference. These reference standards were established at some point in history based on agreement between a group of scientists and government representatives. Temperature is a great example. The standard we use is referenced to the freezing point of water. The units we use should be the most appropriate for the intended purposes. Going between different standards is easy but may cause trouble. Recall the classic mistakes made by NASA scientists using both metric and English units. In the case of the Mars orbiter, two different teams working on the project were using different units, resulting in a $125 million loss (Conradt 2010). In some cases, we may need to work with multiple scales, but it is certainly safer to identify the most convenient scale in each situation and stick with one system.

4.3 VARIABLES

Variables are all those inputs or outputs that we can vary in our experiments. All the inputs and outputs listed in an *Input–Process–Output* diagram could be considered variables. Whether we are varying the inputs/outputs by controlling, measuring, ignoring, or manipulating/managing, it is critical that we understand the roles that each play in our experiments. The goal of scientific experimentation is to examine the relationship between variables. Whether we are attempting to quantify, qualify, establish, study, or determine variable relationships, our experiments will always involve them.

Historically, we have divided variables into two broad categories: dependent and independent. Independent variables are those variables that we control, vary, change, or manipulate in some way while dependent variables are those we measure. In a broad sense, we could think of independent variables as inputs and dependent variables as outputs. From an experimental perspective, the independent variables are those we choose to be not biased by—those that are free from our inputs (the initial conditions) while we monitor our outputs (those variables that depend on the experimental conditions).

As we saw in the Input–Process–Output diagrams, there are many factors that can be inputs, all of which may impact that results or dependent variables. We typically choose one or a few of the input variables to change in our experiments to study the effect or impact on our outputs. What about all the other input factors that we listed? They are still independent, free, unbiased variables, but we aren't intentionally varying them as a part of the experiment. We will divide all our inputs into three categories: constants (C), noise (N), and variables (X) (Ishikawa 1987, Wortman et al. 2007).

C = *Constants:* Constant variables must be held constant and require standard operating procedures to ensure consistency. Some examples of constants are the method used to make a measurement, the method used to load material in a milling or drilling process, and the furnace temperature setting. All variables that we label as "C" should have standard operating procedures written in the process notebook that detail how that variable will be controlled and held constant. Anyone taking data must then be trained to follow the standard operating procedure. Any variability in the output is to a degree a reflection of the variability occurring in the input variables. So the better we control the input variables, the better we control the output variables (our results).

N = *Noise:* Noise (or uncontrolled variables) cannot and will not be held constant. Examples are room temperature or humidity. These variables may or may not impact our results but complete control may be impossible or impractical for our experiment without elaborate and prohibitively expensive modifications to infrastructure. For example, if it's a hot Michigan day and the crappy air conditioning is only sort of working, a university lab where we are attempting to measure the cooling rate of a thermocouple rod may depend on the temperature variations and drafts in the room. The altitude may be a noise factor for an experiment in Colorado, giving different results than if the same experiment were performed in Hawaii. Humidity may affect some experiments such that experiments performed in Arizona and Florida will be different.

X = *Variables:* Variables are key process (or experimental) variables to be tested (varied) in order to determine what effect each has on the outputs and what their optimal settings should be to achieve desired results.

A really good practice is to label each of the inputs on the Input–Process–Output diagram with C for constants, N for noise, and X for variables. For example, writing temperature (C) or humidity (N) immediately informs the reader that temperature is controlled while humidity is not.

4.4 MEASUREMENT = SIGNAL + UNCERTAINTY

No measurement is exact. Absolutely, unequivocally, there is no physical quantity, measurement, constant, or value that can be measured exactly. Every measurement is made up of two parts: the actual signal (magnitude) and some uncertainty (reliability). This is an important point and one that we really do not often think about. However, this concept is critical for our understanding of experimentation as scientists or engineers. Unless we are simply counting using exact integers, we see that repeated measurements of the same quantity give different values. Uncertainty is a part of any measurement. Any physical measurement is composed of two parts:

1. Signal: A number that estimates the magnitude of the effect being measured, and
2. Uncertainty: A number that represents the degree of trustworthiness or reliability associated with the measurement.

Before going any further, I want to establish a convention for talking about uncertainty. In the previous section, I introduced the idea of noise as an effect that we choose *not* to control in our experiment. In other texts, we may see signal and noise defined or discussion of a signal-to-noise ratio, but I want to keep "noise" as a distinct concept. Uncertainty does include noise as it is defined in the prior section. However, uncertainty also includes measurement error, random error, and other sources of error as well. In order to be consistent, going forward, we will talk about measurements as part signal and part uncertainty. Noise will refer to the uncontrolled effects or treatment factors in our experiment.

This idea manifests itself in everyday examples, from the tape measure at a suit fitting that isn't calibrated precisely, to the office or school clocks that vary slightly (sometimes more than slightly) from room to room, to the beakers in chemistry lab, to the speedometers in our cars. Even physical "constants" are approximated to a finite number of digits. How many digits of the speed of light or gravity do you know off the top of your head? When attempting to measure these constants, even with the most sophisticated instruments known to man, the measured value will include some uncertainty.

The effect that we measure will only be as precise (reliable, trustworthy) as our uncertainty. Therefore, in order to be confident in our measurements, we must be confident of our uncertainty. As scientists and engineers, our job includes not only observing the effect but also quantifying the uncertainty in our experiments. Here's where things get a bit messy and confusing. Although the effect appears to be a fairly consistent concept, our understanding of uncertainty has changed over the years and appears to be still developing. One final repetitive note: measured data "is complete only when accompanied by a quantitative statement of its uncertainty. The uncertainty is required in order to decide if the result is adequate for its intended purpose and to ascertain if it is consistent with other similar results" (NIST 2006).

What do we mean by uncertainty? In a typical report, we might see a measurement expressed in the following expression.

$$m = 50 \pm 5 \text{ g} \tag{4.1}$$

What does this *really* mean? Does the scale or balance being used to make the measurement have an uncertainty of 5 g? Does the scale have an uncertainty of 10 g? Is 5 g the standard deviation or the standard error? Is it the expanded uncertainty, e.g., $\pm 2\sigma$ or $\pm 3\sigma$, standard uncertainty, u, or combined uncertainty, u_C? Is the 5 g expression simply the experimenter's best guess at uncertainty?

It is not obvious what this notation represents, and this makes it difficult for experimenters, much less decision makers such as engineering managers or the marketing department in a company, to compare results. According to University of North Carolina physicist, Professor David Deardorf, "The interpretation of u in $x \pm u$ is not consistent within a field of study, let alone between fields of study, and the meaning is generally not specified" (Deardorf 2016). "The ± format should be avoided whenever possible because it has traditionally been used to indicate an interval corresponding to a high level of confidence and thus may be confused with an expanded uncertainty" (Deardorf 2016). However, this notion is commonly used in most fields of study. In many cases, the ± expression is the expected format for uncertainty, even though no one really knows what it represents. If this is a requirement for a publication, paper, or memo, we should include an explanation of exactly what is meant by the ± format in our work.

This is not a subject that teachers and instructors usually spend a lot of time dealing with. However, in the late 1990s, as metrology (the science of measurement) became more of a science in and of itself, seven organizations of international scientists determined that we needed an *international vocabulary of metrology*. These scientists have provided us with the vocabulary necessary to communicate effectively about our measurements. The conceptual framework is laid out in two documents: The Guide to the Expression of Uncertainty (GUM) and The Vocabulary of Metrology (VIM) (GUM 2009, VIM 2012). These documents and their associated supplements provide a full formal process for uncertainty quantification. The National Institute of Standards and Technology (NIST) Guide to the Expression of Uncertainty document captures the work of the international organizations on uncertainty expression and supplements with examples. The NIST Guide to the Expression of Uncertainty (Taylor and Kuyatt 1994) is free for download and will not be duplicated here. The presentation herein of uncertainty will be a simplification of these concepts and provide at least an introduction to the terms and concepts of uncertainty. I would refer anyone publishing data to rigorously follow the guidelines and definitions presented in these valuable reference documents.

All measurements of physical quantities are subject to uncertainties in the measurements. Inconsistency in the results of repeated measurements arises because variables that can affect the measurement result are difficult to hold constant. Even if the circumstances could be precisely controlled, the result would still have an error associated with it. Uncertainty is certain, whether it comes from the metrology manufacturer or estimation in reading the measured value as a result of the scale (as in a meter stick). Of course, steps can be taken to limit the amount of uncertainty. The first major step is characterization of the various sources of uncertainty followed by minimization of the impact on the experimental results.

4.5 AN UNCERTAIN TRUTH

As we discussed in Section 4.4, measurement is the result of the actual signal we are attempting to measure and any error in the measurement. Our job as scientists and engineers is to discern which is signal and which is uncertainty in scientific data. We talk about the quality of our data by examining the signal magnitude as compared to the uncertainty. We can do two things to increase this ratio:

1. Increase the signal

2. Decrease the uncertainty

We will look at both of these options in this section.

4.5.1 Strengthening the Signal

Increasing the signal or the primary effect begins by examining the data collection process. It is critical to understand the data collection process because if we make a decision based on bad data, we could create a lot of problems. (Think about the Food and Drug Administration approving a new pacemaker device, blood glucose meter, or implant material with incomplete or inaccurate data.) We need to know why we are collecting data. What's the purpose of the data? Here are some common reasons:

- Characterize, optimize, or monitor a process
- Verify or test a hypothesis or theory
- Identify, verify, or analyze a relationship between inputs and outputs

In addition to understanding why we are collecting data, we need to understand what decisions we are trying to make. This will allow us to have an idea of the magnitude of the data collection we need to undertake. It can also help if we determine which graphs we might need.

Verify and validate the data being collected. We expect that the data will tell us something about the system or process we are investigating, but we want to make sure we know what that process is. Were the data collected under the desired conditions? Is the data representative of the process or system under scrutiny?

Once we are convinced that we are collecting the right kind of data, there are a number of different steps we can take to get a stronger signal. Table 4.2 contains six different methods for increasing signal to noise (Slutz and Hess 2016). The suggestions in Table 4.2 must be balanced with practical considerations and available resources. Of course, any experiment that increases in time will increase both cost and therefore resources, even if it is only a minimal increase. Repeated measurements are a good way to determine how accurately we are measuring an effect. We may find that our measurement equipment lacks the sensitivity that we need to accurately quantify the effect we seek. Randomization is great but may require a lengthy setup time of equipment or other experimental processes. Many experiments are too costly to keep repeating. In this

TABLE 4.2

Examples of simple experimental techniques for increasing the signal and decreasing the uncertainty in an experiment

Method of Increasing S/N	Pros	Cons
Repeat a measurement	Determine if measurement system is adequate	May increase time
Repeat an experiment	Decreases bias and noise	Increased time and resources; May increase cost
Randomizing Samples	Decreases bias and noise	May increase cost, time and/or resources
Randomizing Experiments	Decreases bias and noise	May increase cost, time and/or resources
Increase samples size	Decreases bias and noise	Increased time and resources; May increase cost
Add covariates	Decreases bias and noise	Increased times and resources; May increase cost; Potentially more complex to analyze

Source: Slutz, S., Hess, K., Increasing the ability of an experiment to measure an effect, 2016, http://www.sciencebuddies.org/science-fair-projects/top_research-project_signal-to-noise-ratio.shtml.

case, we, as scientists and engineers, must be careful not to repeat experiments unnecessarily.

Hugh Gauch, an agri-scientist at Cornell, published a paper in *American Scientist* in which he calculated the number of repeated experiments required so that the average of the repeats is more accurate than a single experiment (Gauch 2006). Gauch's calculations are plotted in Figure 4.1. We see that repeating an experiment one additional time, the data are more accurate 60.8% of the time. If we repeat the experiment five times, the data are more accurate 73.2% of the time. What about the other 40.2% or 26.8% of the time, respectively? To achieve 90% confidence of success, the experiment would need to be repeated 40 times. To achieve 95% confidence of success, the experiment would need to be repeated 162 times. On the other hand, the good news is that as we increase the number of repeats, we see an improvement, but beyond a certain point, it is no longer worth it. I am not advocating that all experiments be repeated 162 times or even 40 times; this is more to make us aware that replication has its limitations.

The larger the sample size, the more confidence we can have in the results (see Figure 4.2). The more samples of a whole population we test,

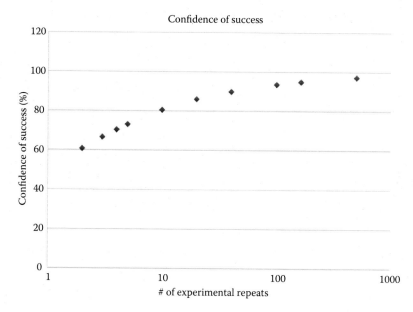

FIGURE 4.1

Law of diminishing returns is seen with confidence in experiments simply due to repeated experiments assuming error in data is all random. (From Gauch, H.G., *Am. Sci.*, March–April, 133–141, 2006.)

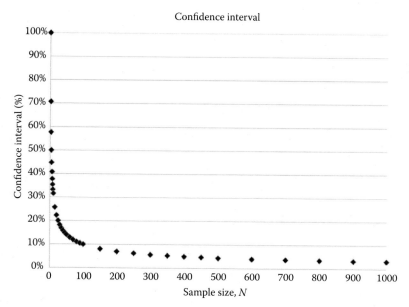

FIGURE 4.2

As sample size increases, the range in the confidence interval decreases. (From Gauch, H.G., *Am. Sci.*, March–April, 133–141, 2006.)

the more representative of the population our results actually are. A good estimate of the confidence interval for a measurement with N samples is

$$C.I. = \frac{1}{\sqrt{N}}. \qquad (4.2)$$

Just think of the number of scientific publications with a single measurement. There is nothing wrong here; this sobering graph should serve to put our experiments in perspective. We write lab reports with single measurements in school, we publish MS and PhD theses with single data measurements and experiments, and in companies, we sometimes make product decisions based on single measurements or experiment. The smaller the effect we expect from our experiment, the more samples we will need (Slutz and Hess 2016).

Background noise (experimental conditions that we don't control—humidity, temperature, etc.) and bias can sometimes have an effect on our experiments. In these cases, by randomizing our experiments, we can reduce the effect of these factors. Randomness can be introduced using statistical software packages or by something as simple as flipping a coin to see which experiment is performed first.

Covariates are factors that vary together to determine the experimental results. In later chapters, we will discuss single variable experimentation and then introduce a more complex type of experimentation, designed experimentation. Single variable experimentation is much easier to understand and model; however, it is impossible to check every set of variables. Therefore, introducing as many covariates into the experiment as possible will provide the most accurate model. There are a number of other important techniques that can be used to increase the signal-to-noise ratio in experiments. However, we'll leave those for advance texts.

4.5.2 Reducing Uncertainty

Let's deal with what is meant by true value and reference value, uncertainty, and error. First, uncertainty tells us the range of values within which the 'true value' can be said to lie within a specified level of confidence. (I'm using quotes around true value because we know there is no such thing.) In order to interpret data correctly and draw valid conclusions, we must indicate uncertainty and deal with it properly. For the result of a measurement to have clear meaning, the value should not consist of the measured value alone. An indication of the uncertainty in the result must also be included. Error is the difference between a measurement and the 'true value' of the measurement (the quantity being measured). Since the 'true value' cannot be absolutely determined, in practice, an accepted reference value is used. The accepted reference value is usually established by repeatedly measuring some NIST or *Bureau International des Poids et Mesures* traceable reference standard. Such reference values are not the "right" answers. Reference values are measurements that have uncertainty associated with them as well and may not be totally representative of the specific sample being measured. In our measurements, we may establish the reference value through repetition.

Now you are probably wondering, what exactly are the sources of uncertainty? For the sake of simplification in this book, we will break uncertainty into three primary categories: *unintentional variation* (mistakes), *random variation*, and *systematic variation*. In the next three chapters, we'll take each of these and examine them closely. Chapter 5 introduces unintentional variation, which includes mistakes. Mistakes are going to happen. We will deal with strategies to minimize mistakes. Known mistakes should never be included in the data presentation. For example, if two lab partners are recording data and one misses a decimal or if recorded data make the value of the measurement ridiculous or physically impossible, never include this in a calculation or presentation. Repeat the measurement; if that's impossible, throw it out, especially if we know it was incorrect. Strategies to minimize opportunities for mistakes are covered in the next chapter. Systematic variation will be limited to uncertainty in the measurement system, which will be covered in Chapter 6. Random variation will be covered in Chapter 7.

The process of determining the uncertainty of a measurement is an extensive process involving the identification of all inputs (all the major process and environmental noise variables) and an evaluation of their effect on the measurement. The most common statistics used to represent uncertainty are range, standard deviation, and standard error. There may be times when results are quoted with two errors. The first error quoted is usually the random error, and the second is the systematic error. If only one error is quoted, it is the combined error.

The total uncertainty (aka combined standard uncertainty) is defined as the combination of all types of uncertainty. The two classes of uncertainty are divided into Types A and B. Type A uncertainty (denoted u_A) is the random uncertainty that is evaluated statistically using either the standard deviation or the standard error. Type B uncertainty (denoted u_B) is based on scientific judgment of all available information about the measurement system (e.g., instrument precision and accuracy, variation in previous data, physical factors, resolution, calibration, etc.). The combined standard uncertainty (denoted u_C) is given by $u_C = \sqrt{u_A^2 + u_B^2}$. Let's look at an example.

Let's say we use a meter stick to make five measurements of a table. The widths measured are 56.2, 56.7, 56.3, 56.9, and 56.5 cm. The standard error yields $u_A = 0.13$ cm. The assumed accuracy and resolution of the meter stick is $u_B = 0.1$ cm. Therefore, we can calculate the combined standard uncertainty.

$$u_C = \sqrt{u_A^2 + u_B^2} = \sqrt{(0.13)^2 + (0.1)^2} = \sqrt{0.027} = 0.16 \text{ cm} \qquad (4.3)$$

The Vocabulary of Metrology and Guide to the Expression of Uncertainty standards also have clean guidelines on how to express uncertainty. In the previous example, the width of the table could be stated as

1. $w = 56.52$ cm with $u_C = 0.16$ cm,
2. $w = 56.52\ (16)$ cm, where the number in parentheses is the numerical value of u_C and refers to the corresponding last digits of the quoted result, or
3. $w = (56.52 \pm 0.16)$ cm, where the number following the symbol $\pm$ is the numerical value of u_C and not a confidence interval.

Type A uncertainty varies in a random, unpredictable way. It is not possible to correct for random fluctuations in the data. This type of uncertainty is calculated using statistical methods. The random component of uncertainty can be characterized and will decrease with repeated measurements. *Random variation* is inherent in measurements (Wortman et al. 2007).

Resolution error from lack of equipment sensitivity or hitting the limit of the device's resolution is a common source of variation. A measurement device may not be able to respond to or indicate a change in some quantity that is too small. This can happen when the smallest division on the device has to be estimated. In general, we agree that the resolution limit is half the smallest division on the instruments scale. For example, a mass balance with the smallest division of 0.001 g will have an uncertainty of ± 0.0005 g. If we measure a part to be 5.446 g, we know that the mass is somewhere between 5.4455 and 5.4465 g. With a single measurement, this expression represents the smallest uncertainty in the measurement.

The resolution limit of the instrument can be thought of as random variation (Type A). However, we also need to consider calibration errors or any offset in the readings that may not be assigned to random variation.

Noise is extraneous disturbances that are unpredictable or random and cannot be completely accounted for. Two common examples of noise that generate random or Type A uncertainty are Johnson–Nyquist noise (also known as thermal noise) and shot noise (also known as current noise). German physicist Walter Schottky first formulated a theory for inherent noisiness of circuits which he attributed to a fluctuation in the current caused by the discrete nature of the electronic charges. In the 1920s, Johnson and Nyquist identified a noise resulting from the thermal vibrations of stationary charge carriers. These two examples of noise are critical to understand when we are selecting a resister or a detector, for example. Johnson–Nyquist noise is the primary type of noise in resistors. Electrons are constantly moving and their movement increases with increased temperature. These electron oscillations are completely random, making the electronic signal inherently noisy. Shot noise is directly proportional to the square root of the signal. Therefore, as the frequency increases in resistors, the current noise decreases. Shot or current noise is also a consideration in photonic applications. Photon or photoelectron detectors are counting quantum particles of light which are discrete units. They are inherently controlled by random statistical fluctuations.

There are times when it is difficult to exactly define the dimensions of an object. For example, it is difficult to determine the ends of a crack when measuring its length. Two people may likely pick two different starting and ending points. Particles are a big deal in the semiconductor industry and medical industry. We specify the number and size of particles that are "allowed" inside our production areas. Dust particles, metal shavings, and viruses found in normal air have a variety of odd shapes and sizes which make them difficult to measure. For example, the Ebola virus has a long tail and then wraps around on one end.

Another component of the combined standard uncertainty is Type B uncertainty. This includes everything that cannot be evaluated through statistical analysis, for example (NIST 2006):

- Previously measured data
- Measurement instrument information including resolution, accuracy, precision, etc., as specified by the manufacturer
- Experience with or knowledge of the measurement instrument (systematic shifts)

- Experience with or knowledge of the material behavior or properties
- Calibration data for the instruments (ANSI 2016)
- Any reference uncertainties assigned from handbooks, etc.

Systematic contributions tend to shift the measurement results to one side of the mean or the other. The offset due to a systematic shift means that all measurements vary in a predictable way. Equipment can never be calibrated perfectly—even if they are the same model, brand, year, etc. When the systematic shift has been quantified and well characterized, measurements can be adjusted in the resulting analysis. Any displacement can be compensated for.

Most systematic shifts can be corrected only when the reference value (such as the value assigned to a calibration or reference specimen) is known. If the source of the systematic shift has been identified, it should be corrected. Correction may be maintenance, replacement of broken or worn parts, software upgrades, or calibration. Some uncertainty may come from allowable tolerances on the equipment from the manufacturer. No part can be duplicated exactly during its fabrication. There will always be a distribution of tolerances with a mean and deviation. These dimensional deviations, which may be perfectly acceptable to the equipment manufacturer, may result in systematic error in measurements. These contributions should be small and will not be identified by repeating the measurements. Even when systematic contributions cannot be eliminated, they should not be considered random.

Variation is what causes values to differ when a measurement is repeated and none of the results can be preferred over the others. Although it is not possible to completely eliminate uncertainty in a measurement, every attempt should be made to ensure that it is controlled and characterized. In all experiments, make sure that all variables are categorized as either control (C), noise (N), or variable (X or Y). Ideally, more effort goes into determining the uncertainty in a measurement than into performing the measurement itself.

We can learn as much or more from understanding, characterizing, and quantifying uncertainty as we can from simply measuring. The standard combined uncertainty measurement is the result of Type A (random variation) and Type B (experiential, material, and equipment fluctuations) caused by component parts and factors related to the entire system. Sometimes, uncertainty limits of all component factors are well known. In this case, uncertainty of simpler systems can be estimated. In more complicated cases, different investigators may not agree on how to combine the uncertainties. In these situations, it is better to do uncertainty analysis with data where all parts of the system are operating simultaneously— only a thorough calibration of the entire system as a unit will resolve the difference.

4.6   KEY TAKEAWAYS

This chapter has laid the foundation for the upcoming chapters. We discussed different required characteristics of data that we will use. We looked at variables. We covered the different types of input variables that give us our output variables. Being able to identify whether a variable is in-control or out-of-control (noise) is an important beginning to experimentation. In Section 4.4, we saw that the measurement data we collect are both signal and uncertainty. There is no such thing as a *true value*. Characterization of uncertainty is important because it tells us how much we can rely on our signal. We can become more confident in our data by strengthening the signal or reducing the uncertainty. Uncertainty is comprised of a number of different components. An uncertainty estimate should address both systematic and random variation. Including uncertainty with measurements is the most appropriate means of expressing the truthfulness of the results.

As we mature as engineers and scientists, so should our experimental sophistication in the characterization of uncertainty. The next three chapters will delve into approaches to minimize uncertainty introduced by *unintentional variation*, *systematic variation*, and *random variation*.

P.S. Test your understanding of the chapter by examining an experimental setup. Create and input–process–output diagram. Label each of the inputs with a C for controlled variables, N for noise variables, and X for process variables. Consider what it would take the move the noise variables to controlled.

REFERENCES

ANSI. 2016. American National Standards website. http://www.ansi.org. National Conference of Standards Laboratories website. http://www.ncsli.org. ANSI/NCSL Z540-2 is a handbook written to assist with calibration laboratories and users of measurement and test equipment.

Conradt, S. 2010. The Quick 6: Six Unit Conversion Disasters. http://mentalfloss.com/article/25845/quick-6-six-unit-conversion-disasters.

Deardorf, D. 2016. Introduction to Measurements % Error Analysis. http://user.physics.unc.edu/~deardorf/uncertainty/UNCguide.html.

Gauch, H. G. 2006. Winning the Accuracy Game. *The American Scientist* March–April: 133–141.

GUM. 2009. Evaluation of Measurement Data—Guide to the Expression of Uncertainty in Measurement. Paris: Bureau International des Poids et Mesures. JCGM 100:2008. http://www.bipm.org/en/publications/guides/gum.html.

Helft, M. 2014. Can Big Data Cure Cancer? *Fortune* 170(2):70–78.

Ishikawa, K. 1987. *Guide to Quality Control.* Tokyo: Asian Productivity Organization.

NIST. 2006. National Institute of Standards and Technology NIST/SEMATECH e-Handbook of Statistical Methods. http://www.itl.nist.gov/div898/handbook/2006.

Slutz, S. and K. Hess. 2016. Increasing the Ability of an Experiment to Measure an Effect. http://www.sciencebuddies.org/science-fair-projects/top_research-project_signal-to-noise-ratio.shtml.

Taylor, B. N. and C. E. Kuyatt. 1994. Guidelines for Evaluating and Expressing the Uncertainty of NIST Measurement Results. Natl. Inst. Stand. Technol. Tech. Note 1297, Washington. http://physics.nist.gov/Pubs/guidelines/outline.html. This is sort of a guide to the Guide to the Expression of Uncertainty in Measurement, GUM. Website: http://physics.nist.gov/cuu/uncertainty/basic.html.

VIM. 2012. International Vocabulary of Metrology—Basic and General Concepts and Associated Terms (VIM 3rd edition). Paris: Bureau International des Poids et Mesures. JCGM 200:2012. http://www.bipm.org/en/publications/guides/vim.html.

Wheelan, C. 2013. *Naked Statistics: Stripping the Dread from the Data.* New York: W. W. Norton & Company.

Wortman, B., W. Richardson, G. Gee, M. Williams, T. Pearson, F. Bensley, J. Patel, J. DeSimone, and D. Carlson. 2007. *The Certified Six Sigma Black Belt Primer.* West Terre Haute, IN: The Quality Council of Indiana.

5

Oops! Unintentional Variation

All men make mistakes, but a good man yields when he knows his course is wrong, and repairs the evil. The only crime is pride.

Sophocles, *Antigone*

No book on experimentation or engineering should ever be written without addressing sources of unintentional variation. Unintentional variation can result from oversight, mistakes, or just plain sloppy experimental practices. In the physical sciences, we feel immune to unintentional variation and many times never even consider it. However, I will try to dissuade you from this line of thinking. In this chapter, I've gathered discussion examples from diverse fields including medicine, genetics, economics, and astronomy to illustrate the lack of understanding prevalent in all fields regarding unintentional variation.

As scientific investigators, developing the ability to spot potential sources of inadvertent variation is not only valuable but also essential to ensuring that any work we present, publish, or share otherwise is repeatable. Earlier, I introduced the idea that some variation in experimentation can be treated statistically. There is another type of variation that cannot be treated statistically: variation due to blunders (aka mistakes, bungles, goof-ups, etc.). These are unintentional, unnecessary, and in many cases completely avoidable types of variation. Just as many of the major complications from surgery are caused by mistakes, most of the unintentional variation in our experiments is due to mistakes. We only need look in the mirror to identify the root cause of this variation—ourselves, the human beings designing and performing the experiments. Unintentional variation, if measured consistently, may contribute to either random or systematic variation. In reality, unintentional variation is neither random nor systematic; it is erratic, inconsistent, and unfortunately, inevitable.

Unintentional variation is as pervasive as the sciences themselves. Even the greatest among us are not exempt from unintentional variation. It is a serious and pervasive issue that affects all scientists regardless of their field of study, education, background, or experience. In the first section, we looked at historical examples of basic mistakes that can make or break an experiment. We'll look at a number of common sources of error that increase unintentional variation. The introduction of variation unintentionally into our experiments is often due to lack of control over the details of our experiment, the environment, and/or unintentional biasing. We will then cover several strategies for reducing unintentional variation and what to do when we know that we have "bad" data. Before leaving the subject of unintentional variation, we'll go where most physical science books don't dare to wander by looking at the role of teleological phenomena on experimentation.

5.1 HISTORY OF MISTAKES

It probably isn't necessary to convince you that new engineers and scientists make mistakes. However, most of us don't realize that even great scientists and engineers make mistakes. Some of the biggest names in science and engineering have made basic beginner mistakes. As a matter of fact, some of the biggest mistakes have led to great breakthroughs in science. Dr. Mario Livio, an astrophysicist and author at Space Telescope Science Institute in Baltimore, Maryland, wrote a book about 12 of these great mistakes in his book *Brilliant Blunders*. Included in the roll call of scientists on this list are the rock stars of science: Charles Darwin, Linus Pauling, Lord Kelvin, and Albert Einstein. According to Dr. Livio, 20 of Einstein's original papers contain mistakes (Livio 2013).

Why spend a whole chapter on mistakes? Here's an example from medicine. In the mid-1980s, Israeli scientists found that an intensive care specialist performs an average of 178 individual tasks each day. These tasks range from administering drugs to suctioning lungs, all of which have some amount of associated risk. The amazing thing is that the doctors and nurses were found to make errors on only 1% of the time. However, this amounted to two errors per day per patient. Of the more than 150,000 deaths each year following surgery, studies repeatedly show that roughly 50% of those deaths and major complications are avoidable (Gawande 2010). This example isn't experimentation in the true sense of the word, but if we take the series of actions or steps that hospital staff take each day and put it in a lab, the parallels become clear.

We might argue that this study is old since it is from the 1980s. Surely, we perform better than this more recently. In 2013, Dr. John James published a review article in the *Journal of Patient Safety* entitled "A New, Evidence-Based Estimate of Patient Harms Associated with Hospital Care" (James 2013). He estimates that there are 440,000 preventable mistakes that contribute to the death of patients each year in hospitals. These 440,000 deaths are roughly one-sixth of all deaths that occur in the United States each year. In other words, a significant number of deaths in hospitals could be avoided with procedural changes to eliminate mistakes. The knowledge exists; however, steps are skipped and mistakes are made. In experimentation, we perform thousands of actions to carefully prepare our work. What if we only had only a 1% error rate? Would our work be repeatable and reproducible?

Example

There are many scientific practices that have documented procedures for how to perform sample preparation. However, in 1951, when scientists around the world were trying to grow cells outside of a human body, little was known about the best way to grow cells. At Johns Hopkins University Hospital, George and Margaret Gey were among these scientists. Many versions of the perfect culture to grow cells were tested. One technique in the Gey lab, developed by Margaret from her days in surgical training, involved chicken bleeding. Margaret worked out the procedure and provided step by step instructions for any researcher who wanted to use it.

Additionally, contamination was an ongoing problem. Bacteria were constantly being introduced into the samples via unwashed hands, breath, dust particles, etc., which killed the cells. Through Margaret's surgical training, she knew the most up to date practices regarding sterility. Like those in the Gey's lab, most scientists working on this problem were biologists who knew nothing about contamination at that time. Margaret taught everyone in the Gey's lab, from her husband George to the lab techs to the graduate students and scientists, about preventing contamination. It is said that she is the "only reason the Gey lab was able to grow cells at all." The cells they eventually grew and shared with labs around the world were from a young mother who died from cervical cancer, Henrietta Lacks. The cells were therefore given the name HeLa, using the first two letters of her first and last names.

Although contamination control was identified as critical in the growth of cells, few realized what a pivotal role contamination would play in the future of cell growth. HeLa was being grown in labs around the world and in space at zero gravity by 1960. Historic discoveries of monumental importance resulted from work with these cells—discoveries such as the carcinogenic effects of cigarettes, x-rays and certain chemicals. HeLa was useful in the discovery of chemotherapy drugs. In labs around the world, researchers began to notice that normal cells eventually became cancerous. All cells eventually behaved like HeLa cells. Soon it was feared that all cultures were contaminated with HeLa cells.

Scientists were lackadaisical with their cultures. Records were not detailed and many were mislabeled or not labeled at all. Research that was cell-specific was soon deemed worthless due to this lack of precision. Although there were calls for improved methods of handling cells, it wasn't until September 1966 that the attention of the scientific community was really gained. The biologists had no idea how hardy these HeLa cells were. They could hitch-hike on dust particles, on unwashed hands, on used pipettes or dishes, clothes, shoes or through ventilation systems. If just one HeLa cell reached a fresh cell sample, the HeLa cells took over. These results (known as the HeLa Bomb) brought millions of dollars of research into question just because scientists were cavalier about the environment in their labs. (Skloot 2010)

These types of examples of unintentional variation are prevalent in all science and engineering fields. Most experiments/problem solving activities are complex multivariate situations. There are many moving pieces. Keep in mind that unintentional variation is difficult to quantify. Therefore, the scientists and engineers solving the problem are ultimately responsible for being aware of the potential sources of variation and controlling each source or accounting for the source in some way.

5.2 UNINTENTIONALLY INTRODUCING VARIATION

Some of the more common sources of variation are (1) uncontrolled environments, (2) inconsistent measurements, (3) different reaction times, and (4) ineffective communication.

The environment in which certain experiments are performed can be a large source of variation. By environment, I mean everything from temperature, humidity, atmospheric pressure, lighting, noise, vibrations, electronic emissions, etc. While working on my PhD, one of my experiments was very sensitive to noise (light noise and vibrational noise); therefore, I collected data in the middle of the night and on weekends when no one else was around. Although the optical setup was on a stable table, the signal from my plasma was very low under certain conditions compared to the noise. The experiments were incredibly complex to set up. The lab was a shared space between several professors at the University of Michigan in the basement of Naval Architecture and Marine Engineering Building. I knew the schedule of most of the other graduate students and would coordinate with them. One Saturday while running my experiment, I was startled when all the overhead lights suddenly came on. I screamed from the back of the lab (roughly the size of a football field). Professor Ron Gilgenbach, now chair of the Nuclear Engineering and Medicine department, had stopped by to check on his experiment. As a result of the shock, both he and I were in full amygdala activation mode. We were both able to laugh about it minutes later. The point of this story is that this type of environmental variation could have resulted in either systematic or random shifts in my data. Other physical examples of similar shifts might be measurements made in different environments where the results are sensitive to vibrations, drafts, humidity, changes in temperature, electronic noise, etc.

Another common source of unintentional variation is inconsistent measurements. Inconsistent measurements could occur when the person making the measurement doesn't calibrate or zero the equipment. Hysteresis may also result in variation if the equipment has some memory effect from the previous measurement. A poor electrical connection may result in errant values on the equipment display. As graduate students, we will often make our own measurements, prepare our samples, and run our own tests. However, occasionally, younger graduate students, undergraduates, or other support staff are involved in some or all of the steps. Unintentional variation can be added to the data if there is a lack of training, skill, or overall physical ability or operations are performed in a different sequence. Misreading the scale divisions on an instrument display, whether this is due to reading the wrong number, miscounting the scale, or the parallax effect, is an example of a case where inconsistent measurements might be introduced. This variation introduced into the results may result from the distance of the person making the measurement from the scale or indicator used in the measurement. Uncertainty might be added by the angle of view while we are making the reading on a burette, pipette, column, or beaker. The parallax effect is measurement variation where the data are collected with our eyes at different angles, which could result in either a systematic shift or random shifts in the results. If we consistently read the scale with our eyes too low, the values we read may be too high and vice versa. The parallax effect is illustrated in Figure 5.1. Therefore, it is essential that everyone working on the experiment performs the work the same way each time. The way in which equipment is operated has a bearing on the quantity, quality, and consistency of the measurements.

Typical reactions times for most people are between 200 and 300 milliseconds. If we want to distinguish times on the order of seconds for our experiments, we need a more consistent and reliable reaction time to events. There are measurements that vary with time. There are times when equipment needs time to warm up, reach equilibrium, or recover from the prior measurement before use. Lag time may also result in inconsistent measurements either made by the same person or between people. For measuring the time between events, we can easily find equipment that provides measurements to the millisecond. If these measurements are performed by hand, the unintentional variation in the measurements will be significant. We can avoid these by moving to digital data acquisition systems. However, within computer data acquisition measurement

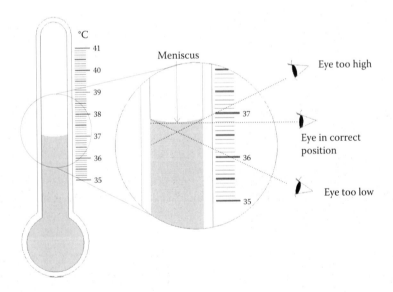

FIGURE 5.1

Illustration of the parallax effect demonstrating how incorrect positions when reading a scale can result in incorrect measurements. Since the molecules of certain liquids are attracted to the sides of the beaker, this surface tension decreases the further away from the sidewalls we get. The surface tension effect creates a concave shape in the liquid with the lowest point is known as the meniscus.

systems, variation is added to measurements that is inherent in the analog-to-digital conversion process. The measurement variation is half of the least significant bit which adds to our systematic variation.

Finally, the most pervasive, most common, and serious source of unintentional variation is ineffective communication. Measurements or actions may be hampered by ineffectual handoff or transitions from one person to another or one measurement to another. Verbal instructions can be forgotten. Written instructions can be misinterpreted. When giving instructions or passing action items off to colleagues or coworkers, other students, or technicians, words like good, bad, tight, loose, uniform, round, tired, full, safe, or unsafe have no precise, specific meaning. In other words, there is a lot of room for interpretation when these adjectives are used in instructions. There are times when we perform actions or steps that seem obvious to us and we completely omit them from our handoff.

5.3   INSURANCE POLICY FOR DATA INTEGRITY

What's the one thing we can do to improve any process? To improve processes, we can minimize variation in the process. One of the best ways to minimize variation is to perform the process (experiment or procedure) the same way each and every time. Almost every process can be improved by eliminating variation (Deming 1982).

In order to accomplish this goal, every lab partner needs to do all the process steps the same way each and every time, independent of who is performing the process. If we are comparing data between experimental groups, then it is especially important that all the data are collected the same way. Comparing data between groups and/or between different people within a group is an important muscle to develop early.

While at the University of Michigan, I was involved in a project that had its roots in the rapidly growing semiconductor industry. Feature sizes continued to shrink, according to Moore's Law, making semiconductor equipment more complex. Semiconductor equipment manufacturers began struggling to ensure that their process chambers performed the same way each time. The equipment manufacturers wanted to assure their customers that the chamber-to-chamber variation and run-to-run variation was small. Plasma physics, although it comprises 99% of the universe, had only been created in a laboratory setting a century earlier. This fourth state of matter was only then beginning to find a lucrative niche with semiconductor etch and deposition technologies. The designers of semiconductor processing equipment thought they were building their process chambers the same way each time, but semiconductor fabrication facilities (affectionately known as fabs) around the world would have problems matching between chambers and between fabs. They even had difficulty matching one run to the next. This was a costly issue for the industry. Intel, Hitachi, Siemens, and other companies instituted "copy exact" worldwide.

In an effort to address this global concern from fabrication facilities and semiconductor equipment manufacturers, researchers from the National Institute of Standards and Technology (NIST), along with academics around the world, designed a tabletop research experimental plasma chamber that could be duplicated at each lab. Plans were drawn up and a total of eight universities and NIST each built the Gaseous Electronics Conference Reference Cell (GEC 2005). The plasma physics community, along with the scientists and engineers at NIST, saw this as an opportunity to develop a deeper understanding of the fundamental interactions between hardware components and the plasma (Brake et al. 1995). Each research team participated in the design of a reference chamber. We each used the same plans to build a plasma tool. Large optical windows were added as viewports to the plasma chamber to allow for optical diagnostic comparisons. The details of what it took to match the hardware such that the process results from chamber-to-chamber required effective communication among the team members. Effective communication was achieved through a variety of venues. Clearly defined specifications for the mechanical and electrical components were required. There were regular meetings, conference calls, and data sharing between groups. These collaborative research efforts resulted in numerous papers, master's and PhD's for many students, and a much clearer picture of just how important every detail was in matching process performance for equipment suppliers. It was through sharing data, plans, and practices that a better understanding of chamber matching became a reality.

Effective communication is probably the single greatest weapon we can wield against *unintentional variation*. Whether we are working with research groups or colleagues around the world or next door, whether we do all the work ourselves or among our research or work group, it is essential that we all perform the work in the same way. The best means for accomplishing this goal is through documentation. Specifications, requirements, protocols, and checklists are all means of clearly defining a set of actions that lead to a repeatable and reproducible experiment free of unintentional variation.

Example

Charles Darwin, credited with the development of the theory of evolution, made amateurish mistakes in his data collection and notes. For a scientist who kept a fairly detailed journal of his travels to the Galapagos Archipelago, he had many glaring omissions of details. He attempted to fill in the details by memory later when he realized that these might be important but his post-travel journal entries could never be verified. I'm referring to Darwin's finch collection from the different islands, although at the time, he didn't realize they were all finches. His purpose in collecting the birds which all looked different—from beak shape and size to feather color—was to send them to John Gould, the head of The Zoological Society and an eminent British ornithologist. In Darwin's Ornithological Notes, he details the location of only 3 of the 31 species he brought back. After studying these birds, Gould concluded they were indeed all finches. Darwin's careless note-taking and sketchy detailed data related to the location of the collected samples could have cost us the evolutionary theory. These finches had evolved to harvest the food on the islands where they were living. Some of the finches harvested seeds for nourishment and other insects. The island terrains varied and the birds evolved to survive. It appears these finches were some of the earliest immigrants as they show the most advanced evolution. Although Darwin was a knowledgeable and experienced taxonomist, he made a similar mistake with tortoises. The vice-governor of the archipelago, Nicholas Lawson, pointed out that "the tortoises differed from the different islands, and that he could with certainty tell from which island any one was brought." As with the finches, Darwin didn't appreciate what he had because the 30 adult tortoises brought on board the ship were eaten and discarded by the ship's crew. This mistake is so obvious and glaring in hindsight, but remember Darwin didn't know what he was looking for at the time of his travels. The lesson we can take from this is to record as exactly and precisely as possible the details and conditions of our experiments. Those things that seem unimportant at the time may actually be the key to a breakthrough. (Sulloway 1982)

5.3.1 Checklists: A Safety Net

After reading Atul Gawande's book *The Checklist Manifesto*, I was sold on the value of creating checklists. We are all prone to take shortcuts, make assumptions, and snap judgments. These lapses in technique can be life or death in some fields but make our experimental results questionable. In this section, I give several examples from Dr. Gawande's book to illustrate the value of a simple checklist.

The cardiac surgeon, Markus Thalmann, who published a story of the three-year-old drowning victim in that small Austrian hospital, had been working in Klagenfurt for six years at the time of the accident. He saw three to five similar cases per year, none of whom survived no matter how hard the emergency team worked. After reviewing the case records, Thalmann attributed preparation as the primary difficulty. These types of cases required a large team of people and a lot of coordination. Thalmann and his colleagues created a checklist. Everyone from the rescue squad to the telephone operator had a checklist. The rescued little girl was the first successful use of the checklist. Even when Thalmann moved on to another hospital, the team in place continued to use the checklist and save lives (Gawande 2010, Thalmann et al. 2001).

Another example from Gawande's book involves a Johns Hopkins intensive care unit that implemented a checklist for one of the simple tasks doctors perform each day: installing a central line. The results were monitored for a year afterward, and the 10-day central line infection rate dropped from 11% to zero. This simple checklist saved the hospital millions of dollars and saved lives.

In a yearlong experiment performed in Karachi, Pakistan, Proctor & Gamble wanted to test the effectiveness of an ingredient in their soap, Safeguard. The experiment's participants were from 25 neighborhoods in and around Karachi. They were provided with a checklist on how to wash their hands. Other literature was provided that discussed when to wash their hands. The study found a 52% reduction in diarrhea, 48% reduction in pneumonia, and a 35% reduction in impetigo over the control groups (Gawande 2010, Luby et al. 2005).

The origin of the checklist dates back to the mid-1930s. Boeing Corporation's Model 299 long-range bomber could fly farther and faster and carry more bombs than other planes could at the time. A large group of military and manufacturing executives watched its inaugural flight stall and crash, killing two of its five passengers. This craft was much more complex to fly than other planes at the time, with the pilot responsible for managing all the additional features. Although the pilot of the aircraft, Air Corps chief of flight testing Major Ployer Hill, was an expert flyer, the crash was deemed to have resulted from "pilot error" and was labeled "too much airplane for one man to fly." While managing the oil–gas mix separately on the each of the plane's four engines, the retractable landing gear, wing flaps, electric trim tabs, and hydraulically controlled constant-speed propellers, the pilot had not released a new locking mechanism on the elevator and rudder controls. Rather than sending pilots back for more training, Boeing decided to create a checklist to deal with the complex details so even an expert would not need to hold it all in memory. With the aid of this checklist, the Model 299 flew 1.8 million miles without one accident. The simple checklist has been honed and refined by Boeing Corporation for all their aircraft. They've perfected both the art of the checklist and the engineering and flying of aircraft. Boeing is the checklist factory (Boorman 2000, 2001, Gawande 2010).

Checklists are not procedures. They are tools with simple steps that get easily missed, but in the case of an airplane or surgery, missing one step can lead to fatal consequences. In the chemical and physical sciences, we aren't necessarily looking for checklists to save lives but to ensure consistency in our experiments. We want to make sure that all the steps that might be crucial to minimizing variation are followed. They are not meant to be detailed operating instructions (see the next section) so that anyone walking in off the street can perform the task. Checklists should be written in the common language of the profession. They are written for experts on a specific task.

If all persons involved in the problem solving study are well trained professionals, a checklist may prove adequate to creating uniformity and eliminating unintentional variation. However, there may be cases where it is essential to have more detail, and in these cases we want to have operating procedures for each step of our experiment.

5.3.2   Standard Operating Procedures

In some cases, checklists are not detailed enough. We need something that will allow for more complex steps. In these cases, we can take a note from a toy manufacturer, Lego, which creates amazing, step-by-step assembly instructions that even a young child can follow. These simple and direct assembly instructions ensure that each completed Lego structure looks identical.

Measurement equipment manufacturers will sometimes provide general guidelines for use, but the operational instructions are not typically tailored to the specific measurements that we are making. Whether we are working on a laboratory experiment for a grade, working on our theses or running an experiment at work, ensuring that the measurements, sample prep, and tests are performed the same way each time is an easy way to eliminate variation with simple *standard operating procedures*.

In university laboratory where equipment is shared, creating a standard operating procedures document is important. The setup may vary, as may the way samples are held, and stabilization times may need to be included. Graduate students will often have junior students assisting with data collection or experimental setup. With a reference document readily available that details how a measurement is performed, unintentional variation (blunders, mistakes, omissions, etc.) can be avoided. Do we still need standard operating procedures if we're planning to make all our own measurements? The answer is yes. The graduate student who follows us may need to duplicate our results. We want to be able to provide detailed instructions of the work we did.

Procedures and/or work instructions provide details and a step-by-step sequence of activities. How-to guides for writing standard operating procedures and work instructions are available in the Reference section at the end of this chapter. Often, labs or companies have their own guidelines for creating procedures. However, here are a few suggestions for creating comprehensive yet coherent operating procedures (Gregory 2016, Texas A&M 2016, Wortman et al. 2007).

- *Be clear* about the primary objective of the procedure to ensure that anyone making the measurements, preparing the samples, or running tests knows what, when, and how to follow the steps. Clearly state when instructions should be followed exactly and when it is okay to be flexible.
- *Be specific* with graphics and words. Break the process into individual steps including any information that ensures that the measurement or sample preparation will be performed in a standard "copy-exact" way. Embedding videos, flow charts, screenshots, or checklists into standard operating procedures is another way to clarify complex steps. These tools can be used to clarify the relationship between different sets.
- *Be brief.* Mark Twain once said, "If I'd had more time, I would have written a shorter letter." It often takes more time to write something concisely, including all the relevant information and using words efficiently and effectively. It's important to invest the time to create easy-to-understand steps. Outlines and lists are a great way to start. Make sure that the steps are in the correct order. I recommend using language specific to our field of science when writing for a university laboratory. Don't dumb it down. If we are going to use a lot of jargon,

acronyms, or symbols, we should just add a legend that explains the symbols, icons, or codes. Unlike writing a user manual for a consumer product, we can assume that the people performing the work have training in the equipment or processes both in university labs as well as companies.

- *Be accessible.* Keep controlled copies of work instructions where the activities are performed. In a company environment, these should be revision-controlled documents (just like parts) stored in an engineering vault. Updates, changes, or improvements should result in a revision roll. Ensuring that the latest version of a procedure or work instruction is used in environments like university labs where revision control is not available may be more difficult. In those cases, keeping the latest standard operating procedures on a network computer might be the best alternative.

- *Test and update.* Get all the people involved who are performing the measurements, running the test, or preparing the samples to test procedures. Some discretion is required in writing work instructions, so that the level of detail is appropriate for the background experience and skills of the personnel who would typically be using them. Similarly for writing procedures and work instructions, the people who perform the activities should be involved in writing the work instruction. The wording and terminology should also match those used by the persons performing the tasks. We should keep our audience in mind! The level of detail and the language may vary greatly depending on the environment and who is reading and following the instructions.

- *Be thorough.* Remember any instructions or standard operating procedures that are incomplete, incorrect, or leave too much room for interpretation open the door for uncontrolled, unintentional variation. As blogger Alyssa Gregory points out, "If you can't get the steps and details down on paper in an easy to understand and intuitive way, you will probably spend a great amount of time and frustration handling support requests and fixing things done incorrectly" (Gregory 2016).

5.3.3 Input–Process–Output Diagrams

A wise investment of time prior to beginning an experiment is mapping out the *Input–Process–Output* diagram and creating a plan to manage the control (C) variables and minimize the impact of noise (N) variables. We have to work twice as hard to eliminate bad data, when taking simple precautions and planning could save us time and money. The time and money seem like nothing compared to the embarrassment of not catching our own mistakes before data are presented, published, and referenced. We also know from the Input–Process–Output diagram that it is critical to our understanding, control, and quantification of experimental variation, whether it is natural/random variation or some type of systematic variation due to some assignable causes. What we see is that all variation can be captured in one of the 6 M's: machines, methods, materials, manpower, measurement, and Mother Nature. The experimental error directly attributable to humans (manpower) can affect all of the other M's (Wortman et al. 2007). One person may make a measurement differently from one reading to the next. Even if the individuals are consistent in making measurements or operating a piece of equipment, there may be variation between people. Even if a particular group or team is consistent in how they perform the experimental activities, unintentional variation can be introduced between groups or teams. This isn't intended to make anyone feel bad or wrong, lazy, or incompetent. We will never completely eliminate unintentional variation. However, once we distinguish and subsequently minimize the human contribution to variation, we can begin to work to address it.

5.4 DYNAMIC MEASUREMENTS

The time sequence of data should be recorded. Record all the information about the collection of the data in addition to the data. Some measurements are time dependent, for example, the stabilization time on a meter or the hysteresis effect. In order to avoid having these effects contribute to unintentional variation in our results, these effects should be characterized and well understood. It is only then that we can create operating procedures that control these known effects. Time-dependent or dynamic measurements often require advanced mathematics (Holman 2001, Coleman and Steele 1999). Advanced texts have thorough coverage of dynamic measurements. In early experimentation, it is best to make every effort to reach a steady state before making a measurement.

5.5 BAD DATA

Unintentional variation can be caused by accidents, carelessness, or improper, poor, or biased technique and may contribute to variation in the experimental results. Misreading and intermittent mechanical malfunction can cause readings well outside of expected random statistical distribution about the mean. One recent example that really hit home was a misplaced decimal point recorded in a database for a sample. The data from the sample were applied to the whole lot of material. These bad values were published internally and resulted in many, many unhappy people. This mistake by the person who recorded the data, and by all the people who used the data without thinking, cost our company thousands and thousands of dollars. No data set should include known mistakes. Values that result from reading the wrong value or making some other mistake should be explained and excluded from the data set. Many times, I've had students in labs and even new engineers deliver reports that include known bad data. The data and analysis reported contain mistakes and yet are presented in engineering meetings. If a reading varies greatly from the true or accepted value, check for unintentional variation (mistakes, blunders, etc.). Poor repeatability and reproducibility (covered in Chapter 6) are also indications of the unintentional variation at work.

- A careful experimenter and/or group of experiments should recognize mistakes and correct them as soon as they are discovered. If the erroneous values fall outside of the known random or systematic variation sources, the first thing to check is for unintentional variation. It is easier to catch these at the time they are made. If not caught at the time of recording, it is often difficult to determine the exact source of the slip-up.
- Replicating all measurements, if possible, is a good way to detect or rule out possible unintentional variation. Repeated measurements give us more confidence in our results. In Chapter 7, we'll discuss an option for handling what appears to be bad data. However, there are many times when repeating experiments is not practical and even prohibitive.
- Review all data sets to detect and remove any data entry errors. (Use objective statistical tests when possible to identify the outliers if they are questionable, covered in Chapter 7.)

- Data entry errors can be avoided with automated data collection, but even automated collection requires the data to be reviewed.
- As we review our data and begin to perform calculations, we must remember to take care with the significant digits. Avoid unnecessary rounding that will reduce measurement sensitivity. Calculations based on the data should include at least one more decimal position than the data point readings. Rounding data will affect the standard deviation in the data but will not impact the mean.

Although no data set should include bad data or mistakes, the removal of data from a data set should not be done in a cavalier fashion. Often, the final conclusions drawn from an experiment can be significantly affected by mistakes. Removing data can give the impression of data "fixing" or result in a missed discovery. There are cases where the data that are unexplainable are actually the most interesting part, as was the case with the discovery of fermion superconductors. In 1975, Bell Labs scientists were studying the magnetic and crystal-field properties of UBe_{13}. In their search for compounds to use with nuclear cooling and nuclear ordering, they measured a superconducting transition at 0.97 K. These results were inconsistent with their expectations. The measurements were thought to be due to contamination of the uranium filament used in the experiment because it didn't fit the expected pattern of temperature-independent susceptibility or magnetic ordering. The experimenters completely missed the discovery of fermion superconductivity (Chu 2011).

━━━━━━━━━━

5.6 ROLE OF INTUITION AND BIAS

It is by intuition that we discover and by logic we prove.

Henri Poincaré

In this section, we discuss the phenomenon of intuition, hunches, bias, and priming in experimentation and problem solving. It's not just sloppy data collection or mistakes that result in unintentional variation. Experiments, mostly from our social science brothers and sisters, continue to show that these nontangible phenomena (intuition, hunches, beliefs, bias, and priming) play a critical role in experimental results, even in what appears to be cut-and-dry results. The governmental institutions who fund much of the physical science research in the United States acknowledge, "There is a growing body of anecdotal evidence, combined with research efforts, that suggests intuition is a critical aspect of how we humans interact with our environment and how, ultimately, we make many of our decisions" (Gregoire 2014).

Some of the leading theories of human behavior currently explain how we make decisions with essentially two "operating systems." These systems are named *System 1* and *System 2*. System 1 operates using our quick, intuitive, and subjective reliance on our subconscious. System 2 is slow, deliberate, analytical, and logical, relying on our consciousness. These "operating systems" use different parts of our brain in making decisions. System 1 operates in the right brain while System 2 operates in the left brain. The right brain contains the limbic (also known as the reptilian) part of the brain. Because System 1 is lazy and makes quick decisions, System 1 associates any new information with our existing knowledge and ways of thinking rather than considering it as something new (Kahneman 2011). As an easy reminder, going forward, I'll refer to System 1 as *Lazy System 1*. It's important to include this so that we can use caution in jumping quickly to decisions with limited information, especially as new scientists and engineers.

5.6.1   Intuition and Hunches

Malcolm Gladwell opens his popular book *Blink: The Power of Thinking without Thinking* with a story that lauds intuition. The story is about a sculpture of a *kouros*, a striding boy. The story describes the reaction of several art experts who had strong instinctual reactions to the sculpture. Their professional intuition told them it was a fake. However, not one could verbalize why they felt it was a fake. This story appears to sing the praises of this almost magical quality of expert intuition (Gladwell 2005). Although Malcolm Gladwell espouses the wonders of intuition and thin-slicing in his best-selling book, I was completely uneasy when I read this anecdote. I firmly believed that the physical sciences were immune to this phenomenon. I was taught that the data and the facts were all that mattered. "Make a habit of discussing a problem on the basis of the data and respecting the facts shown by them" (Ishikawa 1991). Intuition, hunches, and gut feelings were never discussed by physical scientists and engineers, at least around me.

Scientists have, only in very recent history, developed a reputation for being objective. We have worked really hard to establish this reputation as well. However, as I study the great scientists and engineers of history, it is clear that we are human. We are all superstitious. Nobel Prize winner Niels Bohr was once asked about a horseshoe displayed in his home. He responded that it wasn't lucky. "Of course not," Bohr said. "But I understand it's lucky whether you believe in it or not" (Hutson 2015). Isaac Newton wrote as much on alchemy (or more) as he did on calculus (Livio 2013, Weisberg 1993). Johannes Kepler (among others) wrote extensively on astrology (Brackenridge and Rossi 1979). Several well-known scientists refused to consider explanations outside of what appeared to be their own religious mores or personal beliefs, e.g., Albert Einstein and Galileo Galilei (Johnson 2008, Livio 2013). Albert Einstein had a sign posted in his office that read, "Not everything that can be counted counts, and not everything that counts can be counted." Pierre and Marie Curie and Alfred Russell Wallace attended séances (Goldsmith 2005, Weisberg 1993). Newton refused to believe the data because of personal grudges against the scientist who collected the data (Dolnick 2011). Hoyle clung to old ideas while still believing new findings with what appeared to be little discomfort from any chasm that was present (Livio 2013).

Superstitious or intuition-based reasoning still occurs among scientists today. As Matthew Hutson wrote in the *Atlantic* magazine "No one is immune to magical thinking" (Hutson 2015). Dr. Deborah Kelemen and her colleagues at Boston University showed that even physical scientists have a default or fallback bias toward teleology (beliefs) that "tenaciously persists and may have subtle but profound consequences for scientific progress." Dr. Kelemen further explains that even specialized education of physical science cannot break the ties of belief-based explanations (Kelemen et al. 2013).

There are many recent books and articles written on the role of hunches, intuition, and expert or professional judgment in the sciences. Being human, we are designed to rely on our intuition and as trained scientists we are often called on to lean on our professional judgments. However, this can get even the best and brightest into trouble. Lord Kelvin's estimates of the age of the earth were underestimated because he refused to entertain other possibilities outside of his own calculations. Lord Kelvin's big mistake was being "committed to a certain opinion" even though he was "confronted with massive contradictory evidence." Fred Hoyle, one of the greatest astrophysicists in history, similarly remained committed to a steady-state universe even when faced with data from Georges Lamaitre and Edwin Hubble that the universe was expanding (Livio 2013). The Russian chemist Dmitri Ivaovich Mendeleev, who gave us the periodic table of elements, believed that the atom was the smallest particle. According to historian and author Barbara Goldsmith, Mendeleev stubbornly refused to believe Henry G. J. Moseley when he claimed to have discovered a smaller particle, the electron. Goldsmith also wrote of Pierre Curie's antagonistic relationship with fellow physicist Ernest Rutherford. Pierre stubbornly clung to his own theories about radioactive elements. The two scientists aired their dispute publically. Fortunately, and unlike Hoyle and Mendeleev, Curie finally duplicated Rutherford's experiments and conceded (Goldsmith 2005).

We employ intuition and hunches daily. Psychologists believe that "intuition is a rapid-fire, unconscious associating process … The brain makes an observation, scans its files, and matches the observation with existing memories, knowledge, and experiences" (Brown 2010). According to Christof Koch, president and chief scientific officer of the Allen Institute for Brain Sciences, "Intuition arises within a circumscribed cognitive domain. It may take years of training to develop, and it does not easily transfer from one domain of expertise to another" (Koch 2015). Unfortunately, my skills in Scrabble do not transfer to the *New York Times* crossword puzzle. The 1978 Nobel Memorial Prize in Economic Sciences winner Herbert Simon defines intuition: "The situation has provided a cue; this cue has given the expert access to information stored in memory, and the information provides the answer. Intuition is nothing more and nothing less than recognition." Simon's definition "reduces the apparent magic of intuition to the everyday experience of memory" (Simon 1992).

In *Thinking, Fast and Slow*, the 2002 Nobel Memorial Prize in Economic Sciences winner Professor Daniel Kahneman reviews some of the research addressing the "marvels and flaws of intuitive thinking" (Kahneman 2011). "Intuitive answers come to mind quickly and confidently, whether they originate from skills or heuristics." The solution is to "slow down and construct an answer." However, knowing all this doesn't make a difference. He goes on to say, "Except for some effects that I attribute mostly to age, my intuitive thinking is just as prone to overconfidence, extreme predictions and the planning fallacy as it was before I made a study of these issues." We, whether a trained scientist/engineer or armchair scientist/engineer, are forecasting, predicting machines. While driving to work, we are constantly anticipating what the other drivers will do. At sporting events, we see skilled athletes predicting the next move of their opponents. We predict the reactions of our partner or spouse when we deliver bad news. Our quick predictive judgments are based on data from past experiences, Lazy System 1 thinking. As engineers and scientists, we must deliberately and logically use our models and calculations to predict and theorize certain performance based on past experiences or experiments. This is using our knowledge of subject matter to guide our work, System 2 thinking. Dr. Khorasani describes this type of predictive judgment as "the guiding light that helps the researcher" (Khorasani 2016). However, most other predictions use intuition. These judgments are based on skill and expertise or intuitions that are "sometimes subjectively indistinguishable" from skill and expertise but "arise from the operation of heuristics that often substitute an easy question for the harder one that was asked" (Kahneman 2011).

This field of research on the role of intuition and hunches is rife with debate. There remain a number of scholars who value human judgment over algorithms. The scientists who are studying the role of insightful behavior in problem solving have found that intuition is the result of expertise rather than sudden realizations (Lung and Dominowski 1985, Wan et al. 2011). Repeated studies support the accumulation of knowledge in problem solving as the result of a gradual process (Bowers et al. 1990, Weisberg 1993). The hunches of scientists may be dependent on the accumulation of information from the problem. We often deal with new situations on the basis of what we've done in similar situations in the past. Weisberg calls this "continuity of thought." He uses Thomas Edison's development of the kinetoscope as an example of this type of thinking. Edison's invention of the kinetoscope is based on his earlier invention of the phonograph (Weisberg 1993). Professor Weisberg shows that ideas and intuition come from the accumulation and acquisition of information and experiences about the problem we are attempting to solve. Solutions don't appear out of the blue but hit us a like a lightning bolt or magic wand.

Psychologist Paul Meehl analyzed studies of clinical predictions based on subjective impressions from trained professionals (professions where judgment is required at work). In one study, he found that statistical algorithms were more accurate than 11 of the 14 counselors. The number of similar studies comparing algorithms to humans continues to grow. In 60% of roughly 200 studies, the algorithm is more accurate. The remainder of the studies found a tie between humans and algorithms (Meehl 1986).

As new scientists and engineers, we will develop our subjective skills in time as we gain experience. However, we must be cautious and rely on objective data as much as possible. There are no shortcuts; those of us in the physical sciences will need to wait it out and base our recommendations on data. In the meantime, just to be on the safe side, there are a couple of things to avoid.

- Avoid any emotional bias, beliefs, or preconceived notions about what is supposed to be or happen. Let the experimental results speak for themselves, even if they disagree or appear to disagree with other results or our own intuition.
- Count, measure, record data from displays as they are. We want to avoid removing the data based purely on a hunch. There are statistical tests that can be used to identify outliers to our data. This is a much more objective approach. Repeated outliers or anomalous data beg further investigation.
- Use statistical or mathematical algorithms to explain hunches. Do your best to confirm the results by repetition. Remember Richard Feynman's admonition, "The first principle is that you must not fool yourself—and you are the easiest person to fool" (Feynman 1985).

As new scientists and engineers, we want to rely on data and facts and let our intuition develop.

5.6.2 Paradigms

In addition to personal beliefs and humanness, Professor Thomas Kuhn, in his classic text *The Structure of Scientific Revolutions*, identifies scientific paradigms that also limit our ability to even see anomalous results (Kuhn 1962). Kuhn defines normal science as "research firmly based upon one or more past scientific achievements, achievements that some particular scientific community acknowledges for a time as supplying the foundation for its further practices." Normal science defines our paradigm today. Normal science is different for us today than it was for Aristotle or for Galileo or for Isaac Newton or for Benjamin Franklin or for Marie Curie. Where Roentgen saw x-rays, Lord Kelvin saw an "elaborate hoax." Where Antoine-Laurent Lavoisier saw oxygen, Joseph Priestley saw dephlogisticated air. Where Newton saw light as material corpuscles, we now see light as photons with characteristics of both waves and particles (Kuhn 1962). In Kuhn's definition, normal science limits what we can see, as these examples show. Just as these great scientists and engineers from history operated within a certain paradigm, we do as well. It is important that we acknowledge our own System 1 thinking—the paradigms, assumptions, rules, ideas, thoughts, and prejudices that could limit our contribution.

5.6.3 Bias and Priming

There are two additional concepts that may result in unintentional variation in an experiment: priming and cognitive bias. Priming can be loosely defined as cognitive association, while cognitive bias is decision making based on a subjective social reality rather than objective data. Both priming and bias play an important role in all experimentation and can alter the results we get in a number of ways. In *Lean In*, Facebook Chief Financial Officer Sheryl Sandberg summarizes an example of priming from the results of a number of researchers: "When girls were reminded of their gender before a math or science test, even by something as simple as checking off an M or F box at the top of the test, they perform worse." By reminding the girls that they were girls before taking an objective exam, they were primed with the stereotype that "girls aren't good at math and science." Performance in math or science may appear a straightforward, objective measurement; however, unknown biases and priming can influence the results (Danaher 2008, Sandberg 2013). Professor Kahneman has performed repeated experiments examining this effect. In one such study, he showed participants a list of words. By showing words related to money, he created money-primed participants. These money-primed participants were more independent and persistent in problem solving but unwilling to help other participants solve problems (Kahneman 2011).

The point of introducing the idea of cognitive bias and priming is simply to bring awareness. Research into cognitive biases is active, with many researchers working to understand its influence in our lives, our work, and even in the sciences. Economists and psychologists seek to explain how we (human beings) can be manipulated by retailers, politicians, religion, etc. (Kahneman 2011, Willard and Norenzayan 2013). Not even the physical sciences are exempt from biases. In Professor Deborah Kelemen's work, which includes 80 actively publishing PhD physical scientists (chemistry, geoscience, and physics) at Boston University, Brown, Columbia, Harvard, MIT, and Yale, she concludes, "The presence of underlying teleological bias may therefore have subtle enduring effects on our species' intellectual progress, creating impediments for truly mechanistic understanding and discovery even among experts most expected to advance scientific knowledge of nature" (Kelemen et al. 2013). Teleological bias implies that we explain a phenomenon by its purpose rather than a causal relationship.

Correcting intuitive predictions and being cautious to avoid cognitive bias and priming in experimental investigations and research are tasks for System 2 thinking. "The effort is justified only when the stakes are high and when you are particularly keen not to make mistakes. … A characteristic of unbiased predictions is that they permit the prediction of rare or extreme events only when the information is very good" (Kahneman 2011).

In order to overcome our Lazy System 1 thinking (our assumptions, judgements, conclusion, opinions, beliefs, etc.) and shine the light of evidence and data on these, we will need processes, tools, and checklists (Hess 2014).

The following is an example from medicine. Certainly, in recorded history and possibly before that, anyone delivering a baby knew that if a baby wasn't breathing properly within minutes of delivery, there was a high risk of brain damage. "Expert judgment" was historically the method used for determining whether a baby was at risk. Individual obstetricians, physicians, and midwives had their own distress symptoms they looked for. In 1953, anesthesiologist Virginia Apgar created a list of five variables to systematically assess a newborn's vulnerability. The variables were heart rate, respiration, reflex, muscle tone, and color. Each variable had three possible scores. This formula allowed anyone in a delivery room to create a consistent, standardized reference score for each newborn (Apgar 1953). Apgar's algorithm is credited with saving the lives of thousands of babies and is still in use in delivery rooms today (Finster and Wood 2005). As this example illustrates, reliance on our Lazy System 1 thinking can limit our progress rather than allow us to discover for ourselves the world around us.

Given our goal of minimizing unintentional variation in our problem solving and experimentation, it is necessary to understand the filters, blinders, and prejudices that we bring to our work. Only then can we operate with our System 2 thinking.

5.7 KEY TAKEAWAYS

Unintentional variation will happen in experiments. Therefore, a solid experimental protocol is a good insurance policy against mistakes. While there are "no perfect stories," mistakes can be minimized with good lab practices, maintenance, inspection, training, and robust experimental planning. Create checklists, operating procedures, and work instructions to minimize variation. The primary purpose of a written protocol is to minimize variation. If we are doing the setup, running the experiment and making the measurements ourselves, having a detailed procedure may not be necessary. However, I'd recommend it anyway. If we have details of exactly what we did at each step documented, we can always retrace steps if the experiment needs to be reproduced at a later time. The procedures could be somewhat generic, even covering several processes in text or graphic format. The detail in each discrete step will vary depending on the situation, but we can include information about defects to avoid, safety hazards or precautions, required tooling or consumables, and any information that ensures the process will be performed in a standard way. We may choose to describe the process at a general level or provide details and a step-by-step sequence of activities. Flow charts may also be useful to show relationships of process steps.

In this chapter, we also looked at the inadvertent effect that intuition, beliefs, bias, and priming can have on our experiments. The topics are fairly new to physical scientists and engineers, and therefore, we should keep our eyes open to possible variation introduced from these phenomena.

P.S. Try creating a standard operating procedure for a piece of metrology equipment. Have several people try out your procedure. If possible, have both experienced and inexperienced people perform the procedure. What did you learn? Was it difficult to write? Did you need to make improvements?

REFERENCES

Apgar, V. 1953. A Proposal for a New Method of Evaluation of the Newborn Infant. *Current Researches in Anesthesia and Analgesia* 32:260–267.

Bowers, K., G. Regehr, C. Balthazard, and K. Parker. 1990. Intuition in the Context of Discovery. *Cognitive Psychology* 22:72–110.

Brackenridge, J. B. and M. A. Rossi. 1979. Johannes Kepler's On the More Certain Fundamentals of Astrology, Prague 1601. *Proceedings of the American Philosophical Society* 123(2):85–116.

Brake, M. L., J. T. P. Pender, M. J. Buie, A. Ricci, and J. Soniker. 1995. Reactive Ion Etching in the Gaseous Electronics Conference RF Reference Cell. *Journal of Research of the National Institute of Standards and Technology* 100(4):1995.

Brown, B. 2010. *The Gifts of Imperfection: Your Guide to a Wholehearted Life*. Center City, MN: Hazelden Publishing.

Chu, C. W. 2011. The Evolution of HTS: T_c-Experiment Perspectives. *BCS: 50 Years* ed. L. N. Cooper and D. È. Feldman. Hackensack, NJ: World Scientific.

Coleman, H. W. and W. G. Steele. 1999. *Experimentation and Uncertainty Analysis for Engineers*. 2nd Ed. New York: John Wiley & Sons.

Deming, W. E. 1982. *Out of Crisis*. Cambridge, MA: Massachusetts Institute of Technology, Center for Advanced Engineering Study.

Dolnick, E. 2011. *Clockwork Universe: Isaac Newton, the Royal Society and the Birth of the Modern World*. New York: HarperCollins.

Feynman, R. P. 1985. *Surely You're Joking, Mr. Feynman: Adventures of a Curious Character*. New York: W. W. Norton.

Finster, M. and M. Wood, 2005. The Apgar Score Has Survived the Test of Time. *Anesthesiology* 102:855–857.

Gawande, A. 2010. *Checklist Manifesto: How To Get Things Right*. New York: Metropolitan Books/Henry Holt and Company. References used from Dr. Gawande's book include the following:

Boorman, D. J. 2000. Reducing Flight Crew Errors and Minimizing New Error Modes with Electronic Checklists. *Proceedings of the International Conference on Human Computer Interactions in Aeronautics*. Toulouse: Editions Cepaudes. 57–63.

Boorman, D. J. 2001. Today's Electronic Checklists Reduce Likelihood of Crew Errors and Help Prevent Mishaps. *ICAO Journal* 56:17–20.

Luby, S. P. et al. 2005. Effect of Handwashing on Child Health: A Randomized Controlled Trial. *Lancet* 366:225–233.

Mellinger, P. S. 2004. When the Fortress Went Down. *Air Force Magazine* October. pp. 78–82.

Thalmann, M., N. Trampitsch, M. Haberfellner et al. 2001. Resuscitation in Near Drowning with Extracorporeal Membrane Oxygenation. *Annals of Thoracic Surgery* 72.

GEC. 2005. Gaseous Electronics Conference Radio-Frequency (GEC RF) Reference Cell. *Journal of Research of the National Institute of Standards and Technology* 100(4). The special issue contains the collaborative work of twelve research groups from around the world.

Gladwell, M. 2005. *Blink: The Power of Thinking Without Thinking*. New York: Back Bay Books.

Goldsmith, B. 2005. *Obsessive Genius: The Inner World of Marie Curie*. New York: W. W. Norton.

Gregoire, C. 2014. 10 Things Highly Intuitive People Do Differently. *The Huffington Post* March 19. www.huffingtonpost.com.

Gregory, A. 2016. 7 Tips for Writing an Effective Instruction Manual. http://www.sitepoint .com/7-tips-for-writing-an-effective-instruction-manual/.

Hess, E. D. 2014. *Learn or Die: Using Science to Build a Leading-Edge Learning Organization*. Columbia: Columbia Business School Publishing.

Holman, J. P. 2001. *Experimental Methods for Engineers*. 7th Ed. New York: McGraw Hill Higher Education.

Hutson, M. 2015. The Science of Superstition. *The Atlantic*.

Ishikawa, K. 1991. *Guide to Quality Control*. Japan: Asian Productivity Organization.

James, J. T. 2013. A New, Evidence-based Estimate of Patient Harms Associated with Hospital Care. *Journal of Patient Safety* 9(3):122–128.

Johnson, G. 2008. *The Ten Most Beautiful Experiments*. New York: Alfred A. Knopf.

Kahneman, D. 2011. *Thinking, Fast and Slow*. New York: Farrar, Straus and Giroux.

Kelemen, D., J. Rottman, and R. Seston. 2013. Professional Physical Scientists Display Tenacious Teleological Tendencies: Purpose-Based Reasoning as a Cognitive Default. *Journal of Experimental Psychology: General* 142(4):1074–1083.

Khorasani, F. 2016. The Elements of Effective Investigation. Unpublished.

Koch, C. 2015. Intuition May Reveal Where Expertise Resides in the Brain. *Scientific American*, May/June: 25–26.

Kuhn, T. S. 1962. *The Structure of Scientific Revolutions*. Chicago: The University of Chicago Press.

Livio, M. 2013. *Brilliant Blunders: From Darwin to Einstein—Colossal Mistakes by Great Scientists That Changes Our Understanding of Life and the Universe*. New York: Simon & Schuster.

Lung, C. and R. L. Dominowski. 1985. Effects of Strategy Instructions and Practice on Nine-Dot Problem Solving. *Journal of Experimental Psychology: Learning, Memory and Cognition* 11(4):804–811.

Meehl, P. 1986. Causes and Effects of My Disturbing Little Book. *Journal of Personality Assessment* 50:370–375.

Sandberg, S. 2013. *Lean In: Women, Work and The Will to Lead*. New York: Alfred A. Knopf. Reference contained therein:

Danaher, K. and C. S. Cradall. 2008. Stereotype Threat in Applied Settings Re-Examined. *Journal of Applied Social Psychology* 39(6):1639–1655.

Simon, H. A. 1992. What is an Explanation of Behavior? *Psychological Science* 3:150–161.

Skloot, R. 2010. *The Immortal Life of Henrietta Lacks*. New York: Crown Publishing House/Random House.

Sulloway, F. J. 1982. Darwin and His Finches: The Evolution of a Legend. *Journal of History of Biology* 15(1):1–53.

Texas A&M website. 2016. Guide to Writing Standard Operating Procedures. http://oes.tamu.edu/new/templates/.../SOPs_How_to_Write.pdf.

Wan, X., H. Nakatani, K. Ueno, T. Asamizuya, U. Chen, and K. Tanaka. 2011. The Neural Basis of Intuitive Best Next-Move Generation in Board Game Experts. *Science* 331: 341–346.

Weisberg, R. W. 1993. *Creativity: Beyond the Myth of Genius*. New York: W. H. Freeman.

Willard, A. K. and A. Norenzayan. 2013. Cognitive biases explain religious belief, paranormal belief, and belief in life's purpose. *Cognition* 129:379–391.

Wortman, B., W. Richardson, G. Gee, M. Williams, T. Pearson, F. Bensley, J. Patel, J. DeSimone, and D. Carlson. 2007. *The Certified Six Sigma Black Belt Primer*. West Terre Haute, IN: The Quality Council of Indiana.

6

What, There Is No Truth?

Measurement is the first step that leads to control and eventually to improvement. If you can't measure something, you can't understand it. If you can't understand it, you can't control it. If you can't control it, you can't improve it.

H. James Harrington

Measurement affects every part of our lives. Each package of food has the amount of food written in multiple units. Every doctor visit, even a hangnail, is another opportunity to be weighed. Each time we drive our cars, we are monitoring multiple measurements, speed, temperature, engine revolutions, battery life or fuel level, etc. The cost of a product can be based on its weight as can the cost of shipping. In the daily news, we hear or read of the conclusions that scientists and/or engineers have drawn as a result of some measurement that has taken place. Whether a measurement is needed for experimentation, development, or manufacturing, the uncertainty inherent in the measurement is a source of variation. It is critical that we be able to separate instrument errors from other experimental errors. In an effort to reduce overall uncertainty, the measurement system variation should be one of the first things characterized. In this chapter, we will begin with establishing a common language for describing the measurement system, then look at standards and calibration and tool matching. We will walk through setup of a *measurement system analysis* and look closely at the analysis portion. Finally, we pull back to the big picture and look at the global issues surrounding measurements.

6.1 MEASUREMENT EVOLUTION

Measurement is considered the hallmark of human intellectual achievement. Evidence of measurement tools dates back to prerecorded history. Certainly, some of our earliest artifacts record examples of the use of a scale for relative measurement of an objects weight. The earliest evidence dates back to 2400 to 1800 BCE in Pakistan's Indus River Valley. In these prebanking days, smooth stone cubes were used in weight measurement in balance scales. In Egypt, the scales and stones were used for gold trade—mining yields, cataloging shipments, etc. No scales have survived or at least been recovered to date but multiple sets of weighing stones have been found. The Egyptian hieroglyphics and murals from that time indicate the widespread use of scales in trade.

Time for most of history has been a vague quantity. Nature provided the early measurements. As the sun moved through the sky, the shadows cast by a sundial provided the time of day. Figure 6.1 is a photograph of a sundial prominently displayed in the courtyard of Heidelberg Castle in Heidelberg, Germany. The Roman numerals show the time. Notice the astrological signs to provide the time of the year. The first mechanical clocks were recorded in the fourteenth century. With the advent of mechanical clocks, our understanding of time advanced to include hours and minutes and seconds. Figure 6.2 is an example of an early clock in a tower at Heidelberg Castle with Roman numerals for the digits. Notice the hands have astronomical references to the moon and the sun.

FIGURE 6.1
Sundial at Heidelberg Castle in Heidelberg Germany.

FIGURE 6.2
Clock on the Clock Tower of Heidelberg Castle.

Although the use of measurement devices dates back to prehistory, the first use of devices for indirect measurement was by Galileo. He was the first to use both the telescope and microscope for measurement of natural phenomena (Mlodinow 2008, Randall 2011). Prior to Galileo, scientists followed the Aristotelian model of scientific investigation—faith or direct observations (Dolnick 2011). Aristotle asked why, Galileo asked how. Galileo developed experiments and constructed experiments to test his hypothesis about how the world worked. Harvard Physics Professor Lisa Randall writes of Galileo, "Good science involves understanding all the factors that might enter into a measurement." Galileo and his contemporaries paved the way for Newton, Hooke, Boyle, Wren, and even modern experimentalists. Although our measurement equipment is much more sophisticated and advanced, Galileo's legacy still affects the way we experiment today. Galileo helped "lay the groundwork for how all scientists work today." Randall continues, "Galileo fully understood the methods and goals of science—the quantitative, predictive, and conceptual framework that tries to describe definite objects" (Randall 2011).

The science of measurement has evolved over the centuries. Our understanding of measurement evolved as it became more important in commercial interactions. With increased global trade and manufacturing, we've seen the need for standardization of measurement practices which has given rise to international standards organizations. Metrology, the science of measurement, is derived from the Greek words *metron* ("measure") and *logos* ("word" or "reason"). The word *metrology* arose from two Greek words that together give us logical measurement (Metrology 2016).

Every piece of data we collect is filtered through a measurement system. Almost always, there is some type of gauge involved, a person or computer, and a procedure or method by which the person or computer collects or interprets measurements. A gauge will consist of a detector, which detects the signal and converts it to a mechanical or electrical form—either digital or analog, a signal modifier such as a filter or amplifier and an indicator which will record or control the resultant signal (Holman 2001). We also know that all measurements are a combination of the actual effect *(true value)* and some uncertainty. The variation in our measurements is most likely some combination of both random and systematic errors. With random causes, we expect that the measurements could be on either side of the actual or *true value.* Systematic sources of variation will shift the measurements such that they are not centered on the actual or *true value.* The systematic variation will shift the measurements to either side, in one direction, of the *true value* and thereby shift the whole distribution of measurements to one side of the "true value." Recall from Section 4.5.2, where we distinguished the types of uncertainty, measurement uncertainty is lumped into Type B (variation not due to random variation). In other words, measurement uncertainty characterizes the range of values within which the *true value* is asserted to lie with some level of confidence.

6.2 PROBLEMS

One accurate measurement is worth a thousand expert opinions.

Grace Hopper

There are actually a number of issues with measurements or rather the way we think about measurements. The first and biggest problem with measurements is that we think of measurements as exact. We accept measurements at face value. We base our analysis, assumptions, and calculations on this idea. We make product and process decisions in industrial settings based on measurements. Exact measurements require an ideal measurement system that would produce the same exact measurements each time and these measurements would agree with a reference standard. In addition, the people making the measurements would be perfect and make perfect measurements. There would be no variance in this ideal measurement system. Unfortunately, this utopian measurement system doesn't exist. There is no such thing as a "black box" measurement system.

Second, for most of us, when we think of our studies in physics and chemistry measurements, we think of very stable definitions and well-known concepts and ideas, long ago established. Metrology, the science of measurement, is young and continues to evolve. In 1982, the Automotive Industry Action Group pushed for standards in measurement with the creation of the *Measurement System Analysis Reference Manual,* an internationally recognized standard for measurement system evaluation used by retailers, automakers, manufacturers, service providers, academia, and government. One of their objectives was to improve the quality of measurements we make in order to improve the quality of the decisions we make. However, the *Measurement System Analysis Reference Manual* definitions of equipment variation and appraiser variation have both changed in the 2010 versions of the manual (AIAG 2010). If we were to review the same example used in the *Measurement System Analysis Reference Manual* from 1998 to 2003 to 2010, we would see different values calculated for variation. Our definitions used to evaluate the quality of our measurements are still evolving. This requires that we are diligent and deliberate in our measurements. I mention these only to alert you to this situation and warn you that the exact definitions provided herein may change again in the future.

This leads to the next concern in this field of measurement science, which is that two of the leading international groups in the field of metrology, International Standards Organization's *International Vocabulary of Metrology* and Automotive Industry Action Group's *Measurement System Reference Manual,* do not agree on basic definitions, beginning with measurement system itself (AIAG 2010, ISO 2010, VIM 2012). In the future, we may possibly see International Standards Organization's *International Vocabulary of Metrology* used as a reference base on which to build the Automotive Industry Action Group's *Measurement System Reference Manual.* However, until then, we have different standards that we must choose between.

Another problem in the field of measurements science in our standards, the Automotive Industry Action Group's *Measurement System Analysis Reference Manual*, is the terminology. In everyday parlance, we discuss accuracy and precision. We might even use these interchangeably. Other linguistic concepts we use to characterize a measurement system in everyday language are linearity, sensitivity, bias, repeatability, reproducibility, and trueness, just to name a few. In order to be thorough with our experiment, we need to ensure that we fully understand the definitions of these words. We will be tempted to fall back on our intuitive or everyday definitions of these concepts. I've even found myself doing just that as I write this book.

The final concern that I want to highlight here is the fact that there is no such quantity as a *true value*. We get to a solid reference value by averaging the values of repeated measurements. We intuitively know that the "true" value is the actual value that we should get when we make a measurement. However, this elusive *true value* is "unknown and unknowable" (AIAG 2010). University of Colorado Physics Professor John R. Taylor writes "... no measurement can exactly determine the true value of any continuous variable, whether the true value of such a quantity exists is not even clear" (Taylor 1982). We may at times need to speak and write as if there is a *true value* for every physical quantity that we attempt to measure, but it's important we understand what's really going on.

To summarize the current state of metrology in the world today, the measurements we make aren't exact, the definitions continue to change, leaders in the science of metrology don't agree on definitions, and measurement terminology is confusing. These factors combine to make the landscape of measurement science tricky to navigate. However, I argue that it is essential that we scientists and engineers keep ourselves fully appraised of the current climate in an effort to publish, present, or otherwise communicate the best measurements we can possibly make. Definitions are in order.

6.3 DEFINITIONS

In this section, we will review what we know or think we know about measurement terminology since there is a lot of confusion surrounding the definitions. The first step in communicating the results of a measurement or group of measurements is to understand the measurement terminology. Using the proper terminology is key to ensuring that results are properly communicated. It can be confusing which is partly due to some of the terminology having subtle differences and partly due to the terminology being used wrongly and inconsistently. (By the way, I recommend that all new scientists and engineers get a copy of the *Measurement System Analysis Reference Manual* and read, highlight, underline, and mark up this document until it is second nature.)

First, let's get clear on the distinction between *gauge* and *measurement system*. The *gauge* is any device used to make measurements; while the measurement system is everything that goes into the measurement including the *gauge*, the data acquisition system, the procedures, the people, etc. Since the measurement system includes not only the measurement device but also the method of measurement and people using the device, we expect there to be some variation or a pattern of variation. In a measurement system analysis, we want to quantify or at least identify/characterize that pattern of variation.

Before considering a gauge study or measurement system study, it is important to confirm that the gauge is sensitive enough for the intended application. There are a number of terms that are related to instrument sensitivity that we should review. These include discrimination, effective resolution, least count, and instrument limit of error. Again, these may have intuitive definitions that may not necessarily be aligned with the precise definitions in the *Measurement System Analysis Reference Manual* or the *International Vocabulary of Metrology*. Both least count and instrument limit of error are concepts used when attempting to estimate gauge uncertainty contributions to Type B; however, neither is used in the *Measurement System Analysis Reference Manual*. The least count on a measurement device is the smallest division marked on the instrument, while the instrument limit of error is the precision an instrument can provide. The instrument limit of error must always be less than or equal to the least count and is usually assumed to be the least count or some fraction of the least count (1/2, 1/3, 1/5, 1/10, etc.). In some cases, the instrument limit of error is provided and in others we may need to estimate it. For example, a resistor may be specified as having a tolerance of 5%, which means that the instrument limit of error is 5% of the resistor's value. In cases where we need to estimate the instrument limit of error, practical judgment should be used. If the scale divisions are large, we may feel comfortable estimating to 1/5 or

1/10 of the least count. However, if the scale is small, we may only feel confident that we can estimate it to the nearest 1/2 of the least count. In still other cases, we may not feel confident in estimating to anything less than the least count. When selecting a gauge for use in an experiment or process, the *Rule of Ten* should be used. This rule states that the smallest increment of measurement for the device should be less than or equal to 1/10 of the tolerance. The gauge should be *sensitive* enough to detect differences in measurement as slight at 1/10th of the total tolerance specification or process spread, whichever is smaller. Inadequate *discrimination* will affect both the accuracy and precision of an operator's reported values.

A measured value is meaningless without some statement of its accuracy. *Accuracy* is defined as the closeness of agreement between a measured value and the reference value, in other words, it is the closeness of agreement between the average value obtained from a large series of test results and the measured value. All that exists is a series of measurements. Therefore, this deviation from the reference value is a lack of accuracy. Accuracy is an unbiased reference value and is normally reported as the difference between the average of a number of measurements and the reference value. Checking a micrometer with a gauge block is an example of an accuracy check. Accuracy is an expression of the lack of error and is largely affected by systematic error. From our normal distribution discussions, accuracy is our location variation indication. There are multiple definitions for accuracy, however, and in an effort to avoid confusion, the *Measurement System Analysis Reference Manual* recommends that we avoid using the term accuracy and use bias instead. I will therefore attempt to be consistent with that guidance here as well.

Precision is the closeness of agreement between independent measurements of a quantity under the same conditions. It is a measure of how well a measurement can be made without reference to a theoretical or reference value. The number of divisions on the scale of the measuring device generally affects the consistency of repeated measurements and, therefore, the precision. Since precision is not based on a true value, there is no bias or systematic error in the value, but instead it depends only on the distribution of random errors. The precision of a measurement is usually indicated by the uncertainty or fractional relative uncertainty of a value. Precision is the closeness of agreement between independent measurements. Precision is largely affected by

Not accurate, not precise

Accurate, not precise

Not accurate, precise

Accurate, precise

FIGURE 6.3

Illustration of experimental results demonstrating accuracy and precision as related to random and systematic errors. The bulls-eye in the center represents the "true value" or target value or reference value that we hope to achieve in our experiment.

random error. Relating this back to our normal distribution, precision is the width variation or standard deviation.

Figure 6.3 illustrates the meaning of accuracy and precision. Accuracy and precision are concepts often illustrated with a target. The target looks like a dart board where the center of the dart board represents the *true value* we seek. Figure 6.3 illustrates the relationship between accuracy and precision and random and systematic uncertainty. These should make the definitions of these terms much more understandable in the way we normally consider them. Recall our problem, though: there is no such thing as "the truth" or the *true value*. This makes hitting that target more difficult because the target doesn't exist. In most experimental scenarios, we do not have a reference value that we can use as a target. As experimenters, the situation depicted in Figure 6.4 is much closer to reality in many real-life situations. As Figure 6.4 shows, we cannot tell the difference between which is more accurate. In other words, without a reference value, there is no measure of accuracy. Remember, the measurements we make include uncertainty in the measurements themselves as well as uncertainty in the quantity we are trying to measure.

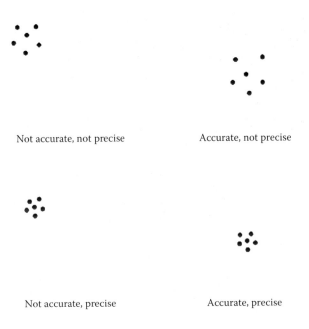

Not accurate, not precise   Accurate, not precise

Not accurate, precise   Accurate, precise

FIGURE 6.4

Sketch of experimental results from Figure 6.3 but without the target. Since we do not know "true value," this corresponds to our experimental situation most of the time.

Most of us learned in school that gravity is 9.8 m/sec² and the speed of light is 3.0×10^{10} cm/sec. However, there is no "true" value for gravity or the speed of light as in Figure 6.5. Figure 6.5a and b show the different measurements of the speed of light over the years. The results are dependent on the method used in the experiment. For the speed of light measurements, the initial measurements were astronomical, a rotating wheel, then a mirror allowed more consistent measurements with reduced uncertainty. Most recently, microwave interferometry was used for the accepted value today of 299,792 km/sec. "If two methods of measuring the speed of light, or for measuring anything were in statistical control, there might well be differences of scientific importance. On the other hand, if the methods agreed reasonably well, their agreement could be accepted as a master standard for today" (Deming 1982). This is exactly what happened. Today, this value is used to define the meter by the *Bureau International des Poids et Mesures*: "The meter

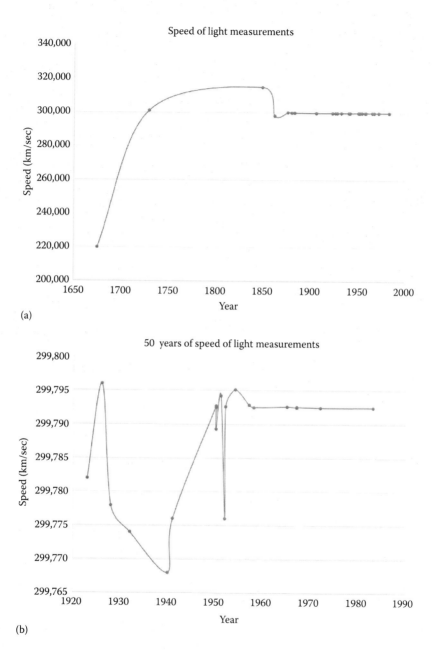

FIGURE 6.5
Measurements of the speed of light from (a) 1675 to 1983 and (b) 1923 to 1983. (Source: Halliday, D., Resnick, R., *Fundamentals of Physics*, John Wiley & Sons, Inc., New York, 1970.)

is the length of the path travelled by light in a vacuum during a time interval of 1/299792458 of a second" (BIPM 1983). This definition has been in place since 1983. Just because a measurement is considered a standard doesn't make it the *true value*.

6.4 MEASUREMENT SYSTEM

There are five characteristics of a measurement system that we are concerned with: bias, stability, linearity, reproducibility, and repeatability. Accuracy is indicated by *bias*, *stability*, and *linearity*, while precision is primarily quantified with *reproducibility* and *repeatability*.

Prior to beginning a measurement system analysis, we should confirm that the instrument will work for the purposes intended. This can be determined with the three indicators of accuracy: *bias*, *linearity*, and *stability*. Bias is the difference between the average value of the large series of measurements and the accepted reference value. Bias is equivalent to the total systematic error in the measurement and a correction to negate the systematic error can be made by adjusting for the bias. Measurements can vary from *true value* either randomly or systematically. *Linearity* describes how consistent the *bias* of the measurement system is over its range of operation. *Stability* describes the ability of a measurement system to produce the same measurement value over time when the same sample is being measured. These three indicators of accuracy are shown in Figures 6.6a and b, 6.7a and b, and 6.8a and b.

In gauge terminology, *repeatability* is often substituted for precision. However, precision cannot be expressed with one value. The precision of a gauge or measurement system describes how "close" the values are to one another. Precision is the random error piece of the measurement system and is represented by the width (standard deviation) in our normal distribution. Precision is expressed with *repeatability* and *reproducibility*.

Repeatability is the precision determined under conditions where the same operator uses the same methods and equipment to make measurements on identical parts. *Repeatability* is the ability to repeat the same measurement by the same operator at or near the same time, as is illustrated in Figure 6.9. In other words, getting consistent results repeatedly

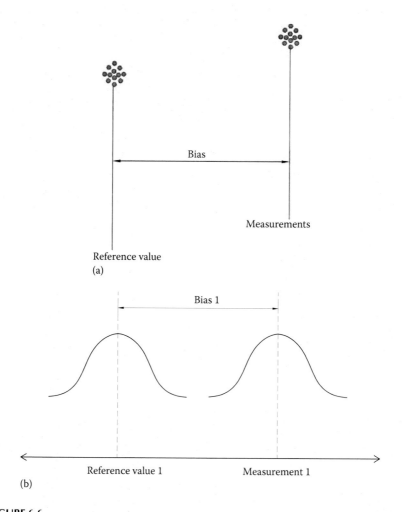

FIGURE 6.6

(a) Sketch illustrating *bias* in a measurement system with data points. (b) Sketch illustrating *bias* in a measurement system with a distribution.

means having the same measurement, same operator, and close to the same time. The *repeatability* contribution to precision is known as the equipment variation.

Reproducibility is the precision determined under conditions where a different operator uses the same methods but different equipment to make measurements on identical specimens. In other words, it is the *reliability* of a gauge system or similar gauge systems to reproduce measurements. The *reproducibility* of a single gauge is customarily checked by comparing the results of different operators taken at different times. Gauge

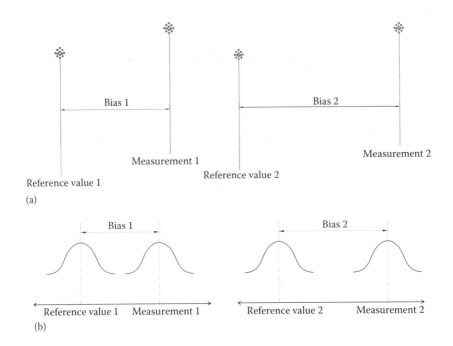

FIGURE 6.7
(a) Sketch illustrating *stability* in a measurement system with data points. (b) Sketch illustrating *stability* in a measurement system with a distribution.

reproducibility affects both accuracy and precision. The *reproducibility* contribution to precision is known as appraiser variation. Appraiser or experimenter variation may be due to the person, operational methods, or the environment. Basically, *reproducibility* measures between-system variation, as illustrated in Figure 6.10.

As an example, let's talk about the phenomenon of *parallax. Parallax*, as we discussed in Chapter 5, is an effect that can create both systematic and random variation in our experiments. When reading an analog meter, the value we see on the meter depends on where we are reading the meter. The amount of liquid in a beaker depends on where we read the scales relative to the beaker increments. These types of tools are meant to be read with our eye exactly in front of the gauge. However, as careful as we try to be, we cannot always position ourselves in the same position each time we read the meter. As a result, there will always be some random uncertainty due to *parallax*. If we ignore the *parallax* effect completely by reading the meter from one side or the other, a systematic uncertainty can be introduced due to *parallax*.

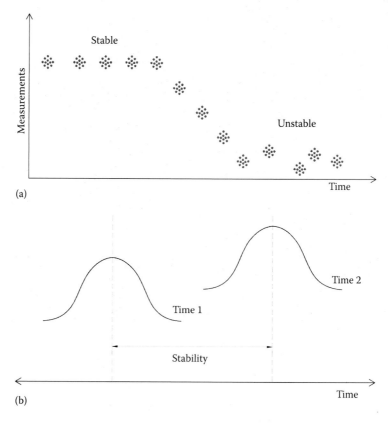

FIGURE 6.8

(a) Sketch illustrating *linearity* in a measurement system with data points. (b) Sketch illustrating *linearity* in a measurement system with a distribution.

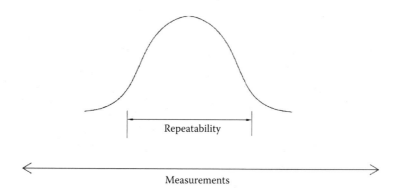

FIGURE 6.9

Sketch illustrating *repeatability* in a measurement system.

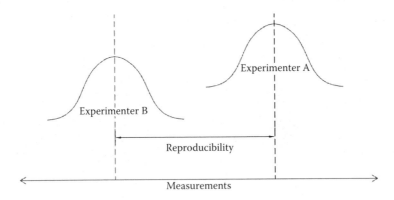

FIGURE 6.10
Sketch illustrating *reproducibility* in a measurement system.

6.5 STANDARDS AND CALIBRATION

As new elements were discovered at the turn of the twentieth century and the science of radioactivity was in the early stages, Marie Curie knew that measurement of radioactivity was critical. She called her challenge "the chemistry of the invisible," and her goal was to be able to identify radioactive elements and quantify the energy. With "tenacious and impeccable work" in metrology, she became the best in the world. Her knowledge allowed her to verify Ernest Rutherford's work in atomic physics. The Curie laboratory became the preeminent authority on the metrology of radioactivity. The unit of measurement for radioactivity was named the curie by the International Radium Standard Committee (Goldsmith 2005).

The need for measurement standards has been known for many centuries. During the Enlightenment, the need for universal measures became a concern, and "To share research, savants had to resort to swapping paper rulers back and forth in the mail to show the length of the particular 'toise' or yard they were using, a patently ridiculous situation" (Marciano 2014). King Louis XVI adopted the metric system as a replacement for all the "disparate systems that had impeded science and were a frequent cause of dispute among merchants" (Mlodinow 2008). In George Washington's first State of the Union address, he declared that "Uniformity in the Currency, Weights and Measures of the United States is an object of great importance and will, I am persuaded, be duly attended to" (Marciano 2014).

The way we handle random and systematic errors is completely different. We want to minimize and completely characterize any random uncertainty in our experiments via statistical treatment and *standard operating procedures*. Systematic uncertainties are much more difficult to identify and characterize. In order to minimize the potential for these uncertainties, we can make every attempt to avoid known issues. Using the highest-quality instruments is a good preventative measure if that is a possibility. Calibration or use of accepted standards is another cautionary measure. As the science of metrology matures and the important ideas and concepts are distinguished, science and engineering gain increasing respect for "practical" (in other words, boring) considerations like calibration. It is such a critical concept that the International Standards Organization requires all companies to use calibrated equipment in all testing and manufacturing of products.

In our recorded history, we have continued to try and come up with ways to support a common measurement tool that everyone trusted. When the primary currency was food or a precious metal, scales were the most important metrology equipment around. The standardization of measurement equipment allows the world to establish systems for use by all industries. Calibration is the comparison of a measurement standard or instrument of known accuracy with another standard or instrument to detect, correlate, report, or eliminate by adjustment any variation in the accuracy of the item being compared. *The minimization of measurement error is the primary goal of calibration systems.* Calibration of measurement equipment is necessary to maintain accuracy but does not necessarily increase precision. In order to improve the accuracy and precision of a measurement process, the process must have a defined test method and must be statistically stable. In laboratories around the world, we will find equipment that has been calibrated by an outside organization. Understanding the standard to which the instrument is calibrated gives experimenters confidence in their measurements.

6.6 MEASUREMENT MATCHING

There may be situations where we need to use two different measurement tools to perform our experiments. These metrology tools may come from the same manufacturer and may even have sequential serial numbers. However, it is almost certain that these two tools will not perform exactly the same and will therefore give different results. It isn't important that these tools give exactly the same results; however, it is critically important that we know how far the results depart from one another. It is important that the equipment is in statistical control.

The semiconductor industry, with metrology tools costing sometimes more than ten million US dollars, has faced tremendous pressure to have their tools perform exactly the same. In order for the measurement to be performed exactly the same, the robotic system handling the wafer needs to be exacting. Tool matching, not only from run-to-run but also between tools, is critical. One example of all that goes into matching metrology tools is described by Dr. Clive Hayzelden. Dr. Hayzelden describes the matching process between two ellipsometers. The process involves measuring a film (typically an oxide film) thickness in five or more locations using the same silicon wafer every eight hours over a five-day period. Both dynamic (cycling the wafer in and out of the tool) and static repeatability tests are performed. Static repeatability provides measurement-to-measurement variation, while dynamic repeatability captures the variation in robotic accuracy and focus. The stability of the measurement tools is determined from performing the same measurement over time. How well the tools match one another is determined by comparing the mean and standard deviation for each measurement site (Hayzelden 2005).

6.7 ANALYSIS METHODS

There are three methods that can be used to quantify error in a measurement system: the range method, the average and range method, and the analysis of variance method, often referred to as ANOVA. Table 6.1 compares the three methods for measurement system analyses. The most accurate method is the analysis of variance method because it allows for the quantification of repeatability, reproducibility, part variation, person variation, and the interaction between the part and people variation. Although the calculations in a measurement system analysis involve only simple mathematical functions (addition, subtraction, multiplication, and division) for each of the methods, the analysis of variance method is the most complex.

Let's walk through the basis for the measurement system analysis model. With anything we are measuring, whether it's a part dimension

TABLE 6.1

Comparison of Three Different Methods of Measurement System Analysis

Method	Pros	Cons
Range (R)	Easy and quick approximation of measurement variability	Cannot distinguish repeatability and reproducibility
Average and range (X and R)	Provides information about causes of measurement error; Able to distinguish between repeatability and reproducibility	Cannot distinguish any interaction between repeatability and reproducibility
Analysis of variance (ANOVA)	Most accurate	Computationally more difficult; typically performed with computer

or the property of a material or an effect of some part changing another part, there is some target or reference value. In the past, and for now, we might have even called this the *true value*. We also know that each part or property that we measure has a certain amount of variability. This means that we can write an equation describing the relationship between the *true value* and variability of a measurement as

$$Reference\ Value = Mean\ Value + Within\text{-}Part\ Variation \qquad (6.1)$$

The *Mean Value* is considered constant while the part-to-part variation is captured in the *Within-Part Variation* term. When we make a measurement there will be some uncertainty in the measurement due to the measurement system and that can be expressed as

$$Measurement\ Uncertainty = Bias + Reproducibility + Repeatability$$
$$(6.2)$$

The *Bias* is a systematic and constant contribution from the gauge, *Reproducibility* is variation introduced by people making the measurements, and *Repeatability* is variation due to repeated measurements using the same gauge and same people. We also know that the

$$Measured\ Value = Reference\ Value + Measurement\ Error \qquad (6.3)$$

When all of these are combined, we get:

$$Measured\ Value = (Mean\ Value + Bias) + Within\text{-}Part\ Variation$$
$$+ Reproducibility + Repeatability \qquad (6.4)$$

The *Mean Value* and *Bias* are placed in parentheses to stress that they cannot be separately distinguished unless a master gauge is used. The contribution to the *Measured Value* from *Within-Part Variation*, *Reproducibility*, and *Repeatability* is random. This model is the basis of the mathematical model used for the development of the analysis of variance method. A detailed development of the mathematical model can be found in the *Measurement System Analysis Reference Manual* (AIAG 2010).

All the methods (and examples presented) ignore the *Within-Part Variation* term (out-of-roundness, out-of-flatness, diametrical taper, etc.) as the data gathering process becomes vastly more cumbersome (AIAG 2010). In order to minimize the impact of the *Within-Part Variation* effect, it is best to capture the maximum within-part variation prior to beginning our measurement system analysis. In addition, confirm that the particular characteristic or property that we are interested in understanding in the measurement system analysis has a much greater effect than the within-part variation.

6.7.1 Setup

The most important part of the measurement system analysis, independent of the method that we select, is the detail of the setup. The measurement system analysis is useful in determining the amount and types of variation in a measurement system and how it performs in its operational environment (as opposed to the manufacturer development lab). We want to allocate the variation to the two categories, repeatability and reproducibility, as we've defined earlier in the chapter. In most practical situations, it is important that we have a well-characterized measurement system (i.e., known bias, repeatability, linearity, reproducibility, and stability) within reasonably established limits.

Before beginning a study of a measurement system, it is essential that we have some information about the tool. From the manufacturer's manuals, we should be able to determine the sensitivity, bias, linearity, etc. However, these are not unique to our individual measurement system nor were the testing conditions equivalent to our laboratory environment.

Recall that Type B uncertainty is all uncertainty not covered by random variation. Therefore, if we do a good job of minimizing any uncertainty due to blunders and keeping our control variables REALLY under control, what is left is primarily excess variation in the measurement system. When we look to reduce variation in our measurements, we need to look at our inputs—all the potential sources of variation. Measurement system analysis is a powerfully simple mathematical tool that allows us to quantify variation due to 3M's: measurement, man (as in human, but two M's and an H don't roll off the tongue in the same way), and materials. After performing a measurement system analysis, we are able to quantify the reproducibility and repeatability of our measurement system. The measurement system analysis allows us to distinguish the variation due to the person(s) performing the measurements. Among the causes listed in the prior chapters, variation due to people could be a lack of training or inadequate/nonexistent checklist or procedures. Whether they become unstable over time or lack homogeneity, a measurement system analysis will allow us to quantify variation due to materials.

In preparation for a measurement system analysis, create an *Input–Process–Output* diagram. Any variation in our experimental/environmental condition that we cannot control becomes a part of the Type B uncertainty. The effects of what we've labeled "Mother Nature" could be any of the following: temperature, humidity, atmospheric pressure, lighting, noise conditions, vibration, electronic emission, etc. We may or may not be able to distinguish or quantify this effect. What influence do the people making the measurements have? Are the procedures for making the measurements well documented? Table 6.2 provides a simple planning tool to assist with setup.

As an added guarantee that the measurements are statistically independent, randomize the order of the parts with each repeat of the measurements. Ideally, the persons performing the measurements would not know the identity of the part they were measuring. The more of a "blind" test we can perform, the more statistically valid our results will be.

TABLE 6.2

Measurement System Analysis Planning Tool

#	Step
1	Create an Input–Process–Output diagram and identify each input with (C) for constant, (N) for noise, or (X) for intentionally varying.
2	Identify how many people will be involved in the study and who they are. Try to select people who normally make or will be making these measurements.
3	Select the sample parts. Determine the number of parts. Label the parts. We will want to select typical parts that are really representative. Remember we are trying to capture the full range of variation that exists within the parts. A good labeling or identification system is important because the parts will be measured multiple times by multiple people. (We may want to try to make this a blind study.)
4	Decide how many times measurements will be repeated. The more critical the dimension, the more measurements we may want to make in order to increase our confidence in the measurements.
5	Ensure that we have adequate sensitivity with our gauge.
6	Confirm that the measurement procedure is well defined and that each person participating in the study is well trained on the procedure.
7	Create a template for logging the measurements. Our template should detail the order of measurements, people, and parts. Stick as close to this template as possible. (This can be done in Excel or other spreadsheet format or using a statistical software package like JMP.)
8	Begin the measurements in the predetermined randomized order of people and parts.

6.7.2 Average and Range Method

Now, let's actually walk step by step through the calculations needed in the measurement system analysis. Table 6.3 provides definitions of the notations used in the *Measurement System Analysis Reference Manual*

TABLE 6.3

Definition of Notation Used in the *Measurement System Analysis Reference Manual*

Symbol	Symbolic Representation
k	Number of people making measurements
r	Number of repeated measurements each person is making
n	Number of parts being measured
$m = r{\cdot}k$	Number of total measurements for each part

calculations. I've summarized the steps in this section, but details and examples can be found in the reference manual.

1. Calculate the average and range (in Excel or spreadsheet per the template), both rows and columns.
2. Calculate the average of the averages for each row and column. This will give us the average of each of the averages of Person A, B, and C's measurements through k and the averages for each of the n parts. To be consistent with the *Measurement System Analysis Reference Manual* template, let $\bar{X}_A$, $\bar{X}_B$, and $\bar{X}_C$ represent the average measurement performed by each of the k measurers: A, B, C, etc., respectively, and $\bar{X}_P$ denote the average measurement over all r^*k measurements for part one through n where $p = 1$ to n for each of the n parts.
3. Calculate the average of the range of all measurements using similar subscripting notation as in Step 2. Let the average of the ranges be $\bar{R}_p$, $\bar{R}_a$, $\bar{R}_b$, $\bar{R}_c$.
4. Compute the average of all the part averages, $\bar{\bar{X}}$, and the average of all the ranges, $\bar{\bar{R}}$, for each of the operators using the following formulas.

$$\bar{\bar{X}} = \frac{\sum_{i=1}^{n} \bar{X}_i}{p} \qquad (6.5)$$

$$\bar{\bar{R}} = \frac{\bar{X}_A + \bar{X}_B + \bar{X}_C + ...}{n} \qquad (6.6)$$

5. Calculate the range of the average for all measurements, $Max\bar{X}$, $Min\bar{X}$, $\bar{X}_{DIFF}$.
6. Calculate the upper control limit (*UCL*) for the range values. The instructions ask us to compare each of the n-part range (R_p for $p = 1$ to 20) values with the upper control limit (*UCL*). Any range that exceeds the upper control limit should be highlighted and examined closely. The highlighted values are significantly different from the others and should be identified and corrected. Once this has been done, the appropriate parts can be remeasured, using the same operator, equipment, etc., as for the original measurements. Then we'd need to recalculate everything.

7. Calculate the following: Equipment Variation (*EV* = *repeatability*), Appraiser Variation (*AV* = *reproducibility*), gauge repeatability and reproducibility (*GRR*), part variation (*PV*), and total variation (*TV*) using the following equations. (See the appendix of the *Measurement System Analysis Reference Manual* for values of K_1, K_2, and K_3. These values can also be found in Professor Acheson Duncan's book *Quality Control and Industrial Statistics*; AIAG 2010, Duncan 1986.)

$$EV = \bar{\bar{R}} * K_1 \tag{6.7}$$

$$AV = \sqrt{(\bar{X}_{Diff} * K_2)^2 - \left(\frac{EV^2}{n*r}\right)} \tag{6.8}$$

$$GRR = \sqrt{EV^2 + AV^2} \tag{6.9}$$

$$PV = R_p * K_3 \tag{6.10}$$

$$TV = \sqrt{GRR^2 + PV^2} \tag{6.11}$$

8. Calculate the percentage contribution to the total variation for each of the metrics in step 7 by dividing each by *TV* and multiplying by 100.

6.7.3 Average and Range Method Analysis

Analysis of the average and range method can easily be performed in a simple spreadsheet or on paper if we are so inclined. The mathematics to compute *EV*, *AV*, *PV*, and *TV* are simple and straightforward.

As we've already stated, *repeatability* can be thought of as a measure of equipment variation. It is the amount of variation in the readings obtained by successive measurements using the same measurement system: methods, tools, etc. *Repeatability* is the variation in the measurements that occurs when the same measurement system is used (equipment, material, method, and appraiser are held constant). *Repeatability* is reflected in the range R_p values. This is our equipment variation, but the individual *R*-average differences may indicate differences in the operators. In this example, R_A is less than R_B and R_C. This tells us that A may have done better at getting the same answer upon repeated measurements of the same part than B or C did. Investigating the difference between A, B, and C's methods might provide an opportunity to reduce variation.

Reproducibility can be thought of as a measure of operator/technician variation. It is the amount of variation in the readings from different measurement systems measuring the same material/parts. This is important because most of the time, in industry and in labs, we have different operators making measurements that are considered the same as other operators' measurements. We could also use *reproducibility* to measure changes in the measurement system. For example, if the same person is making the measurements but using two different methods, the *reproducibility* calculation will show variation due to changes in the methods. *Reproducibility* is the variation that occurs between the overall average measurements for the different operators (appraisers). It is reflected in the $\bar{X}$ values and the $\bar{X}_{Diff}$ value. If, for instance, $\bar{X}_A$ and $\bar{X}_B$ are close and $\bar{X}_C$ is very different, it would appear that C's measurements are biased. We'd have to investigate further to reduce this variation.

Once we have completed the data collection, the next step is to complete the GRR report. The quantity labeled *EV*, equipment variation, is an estimate of the standard deviation of the variation due to repeatability. The quantity labeled *AV*, appraiser variation, is an estimate of the standard deviation of the variation due to reproducibility. The quantity labeled *GRR* is an estimate of the standard deviation of the variation due to the measurement system. The quantity labeled *PV* is an estimate of the standard deviation of the part-to-part variation. The quantity labeled *TV* is an estimate of the standard deviation of the total variation in the study.

If the *GRR* is under 10%, the measurement system is acceptable, and if it is between 10% and 30%, the measurement system may be acceptable depending on how important our work is. If the *GRR%* is more than 30%, the measurement system needs improvement. In this case, the whole process should be examined to determine where the problems are and how they can be corrected. There are many reasons that a measuring system could give erroneous results (variation) (AIAG 2010, Wortman et al. 2007). Table 6.4 shows how these items might appear in terms of repeatability and reproducibility.

TABLE 6.4

Sources of *Repeatability* and *Reproducibility* Error

Sources of Variation	Repeatability	Reproducibility
Part, sample, or material variation	Within part, samples, or material	Between parts, samples, or material
Equipment variation	Within instruments	Between instruments
Standards	Within standards	Between standards
Procedural variation	Within the procedure	Between procedures
Appraiser variation	Within appraiser	Between appraisers
Environment	Within environment	Between environment
Assumptions	Violations of stability and proper operation	Violation of assumptions in the study
Application	Part size, position, observation error	Part size, position, observation error
Software variation	Within an instrument	Between instruments
Laboratory variation	Within laboratory	Between laboratory

There are times when testing is destructive such that it prevents retesting. In these cases, sample or material variation accounts for all the variation within and between samples. Sample variation would account for variation due to form, position, surface finish, or any inconsistency within the sample. Equipment variation can be identified and quantified through measurement system analysis. Equipment variation may show up as a fixed error shift from the *true value* or it may show up with the slow measurement changes over time as with signal drift. Standard variation is unlikely but should be considered. The standards should be more stable than the measurement process. Procedure variation occurs when standard operating procedures are not followed or are not error-proofed. Appraiser variation may occur when one appraiser uses the same gauge and same standard operating procedure but measures variation or it may occur when the measurement system analysis is performed and variation occurs between the different appraisers. Environmental variation may occur within an environment due to short-term changes in the environment or between environments due to differences over time caused by changes in the environment. Examples of environmental factors include temperature, humidity, lighting, cleanliness, etc. Software variation within a program may be a result of variation in the formulas or algorithms, which may result in errors, even with identical inputs or between software versions. Finally, laboratory variation may be a result of variations of measurements within a laboratory or between different laboratories. Testing standards such as those developed by the American Society for Testing and Materials should completely eliminate this risk for the more than 12,500 voluntary consensus standards (ASTM 2016).

6.7.4   Analysis of Variance Method

Random and blind studies are critical with the analysis of variance method to ensure statistical independence. The analysis of variance method is typically performed using a statistical software package like JMP. As a part of this analysis, these packages provide all the graphical and computational analysis. However, interpreting the results and understanding the terms remain the task of the engineer or scientist. Using the graphical and computational results together can provide insights into the data generated with the measurement system.

6.7.5   Measurement System Problems

What if there are problems with the measurement system? If our measurement system is unacceptable at producing repeatable and reliable measurements, this should be addressed prior to beginning our experiments. There are a number of approaches we can take to make improvements. If we have used the measurement system analysis approach, either the range and average method or the analysis of variance method, we can easily identify the areas in our measurement system in the most need of improvement. The *Measurement System Analysis Reference Manual* provides guidelines for systematically eliminating these problems. Once we have addressed the areas of concern, the measurement system analysis should be repeated to ensure that the measurement system is indeed repeatable and reliable.

There are times when the actual physical quantities that we are interested in investigating consist of at least two steps: making a measurement then performing a calculation. Although we'll not get into details of how the uncertainty is propagated into calculations, we need to keep in mind that any uncertainty in measured values will pass through to our calculated values and may be amplified if this is the product or sum of several measured values (Taylor 1982).

6.8   A GLOBAL CONCERN

The study of scientific instruments is so important that there are a number of scientific journals dedicated to instrumentation used in scientific experiments. These journals include *Review of Scientific Instruments, Journal of Scientific Instruments, Instruments and Experimental Techniques, Nuclear Instruments and Methods in Physics, Journal of Astronomical Telescopes, Instruments and Systems, Instruments and Experimental Techniques,* etc. There are eight international organizations that have joined forces to create the Joint Committee for Guides in Metrology. These eight organizations are the International Bureau of Weights and Measures, International Electrotechnical Commission, International Federation of Clinical Chemistry and Laboratory Medicine, the International Standards Organization, International Union of Pure and Applied Chemistry, International Organization of Legal Metrology, and the International Laboratory Accreditation Cooperation. The Joint Committee for Guides in Metrology is responsible for the *Vocabulary of Metrology*. "The mission of *BIPM* (International Bureau of Weights and Measures) is to ensure and promote global comparability of measurements, including providing a coherent international system of units for scientific discovery and innovation, industrial manufacturing and international trade and sustaining the quality of life and the global environment" (BIPM 1983). As we can see, a lot of resources from all over the world dedicate time, energy, and money in addressing measurement concerns. The more we understand the instruments we use from a technical and statistical perspective, the more reliable our experiments will be.

6.9   KEY TAKEAWAYS

Measurements impact all areas of our lives. Our measurements will contain both *random* and *systematic variation*. Measurement systems account for most or at least a large portion of the Type B uncertainty contribution. In order for us to have confidence that we can repeat our experimental results and that our work can be duplicated by other researchers, corporations, or customer groups, we want to ensure that the equipment we use for measurement is properly sensitive, calibrated, and well characterized. A measurement system analysis allows us to do just that. The *Measurement System*

Analysis Reference Manual provides detailed, practical examples of each of these three methods: the range method, the average and range method, and the analysis of variance method. We looked closely at the average and range method as a means of quantifying systematic variation within a measurement system. In Chapter 7, we'll look at quantification of random variation.

REFERENCES

ASTM. 2016. American Society for Testing and Materials International. www.astm.org.

AIAG. 2010. *Measurement System Analysis Reference Manual.* Chrysler Corporation, Ford Motor Company, and General Motors Corporation, 1998, 2003 and 2010. This document is being updated regularly. When performing a gauge study, the latest version should be used as the definitions are honed in the science of metrology.

BIPM. 1983. *Bureau International des Poids et Mesures* (International Bureau of Weights and Measures). www.bipm.org. See Resolution 1 of the 17th CGPM from 1983.

Deming, W. E. 1982. *Out of the Crisis.* Cambridge, MA: Massachusetts Institute of Technology, Center for Advanced Engineering Study.

Dolnick, E. 2011. *Clockwork Universe: Isaac Newton, the Royal Society and the Birth of the Modern World.* New York: HarperCollins.

Duncan, A. J. 1986. *Quality Control and Industrial Statistics.* 5th Ed. Homewood, IL: Irwin.

Goldsmith, B. 2005. *Obsessive Genius: The Inner World of Marie Curie.* New York: W. W. Norton.

Hayzelden, C. 2005. Gate Dielectric Metrology. *Handbook of Silicon Semiconductor Metrology.* ed. Alain C. Diebold. New York: Taylor & Francis.

Holman, J. P. 2001. *Experimental Methods for Engineers.* 7th Ed. New York: McGraw-Hill Higher Education.

ISO (International Standards Organization). 2010. Document ISO/CD 22514-7: Capability and Performance—Part 7: Capability of Measurement Processes. Geneva. http://www.iso.org.

Marciano, J. B. 2014. *Whatever Happened to the Metric System: How America Kept Its Feet.* New York: Bloomsbury.

Metrology. 2016. http://www.french-metrology.com/en/history/history-mesurement.asp.

Mlodinow, L. 2008. *The Drunkard's Walk: How Randomness Rules Our Lives.* New York: Pantheon Books.

Randall, L. 2011. *Knocking on Heaven's Door: How Physics and Scientific Thinking Illuminate the Universe and the Modern World.* New York: HarperCollins.

Taylor, J. R. 1982. *An Introduction to Error Analysis: The Study of Uncertainties in Physical Measurements.* 2nd Ed. Sausalito, CA: University Science Books.

VIM. 2012. *International Vocabulary of Metrology—Basic and General Concepts and Associated Terms.* 3rd Ed. Paris: Bureau International des Poids et Mesures. JCGM 200:2012. http://www.bipm.org/en/publications/guides/vim.html.

Wortman, B., W. Richardson, G. Gee, M. Williams, T. Pearson, F. Bensley, J. Patel, J. DeSimone, and D. Carlson. 2007. *The Certified Six Sigma Black Belt Primer.* West Terre Haute, IN: The Quality Council of Indiana.

7

It's Random, and That's Normal

THE
NORMAL
LAW OF ERROR
STANDS OUT IN THE
EXPERIENCE OF MANKIND
AS ONE OF THE BROADEST
GENERALIZATIONS OF NATURAL
PHILOSOPHY—IT SERVES AS THE
GUIDING INSTRUMENT IN RESEARCHES
IN THE PHYSICAL AND SOCIAL SCIENCES AND
IN MEDICINE AGRICULTURE AND ENGINEERING—
IT IS AN INDISPENSABLE TOOL FOR THE ANALYSIS AND THE
INTERPRETATIONS OF THE BASIC DATA OBTAINED BY
OBSERVATION AND EXPERIMENT

Jack Youden

Earlier in this book, we saw that uncertainty can be broadly divided into *systematic variation* or *random variation*. Systematic variation may be a result of a measurement system or method, but random variation is an inherent part of any measurement. Random variation occurs naturally in nature. No two snowflakes are exactly the same; no two flowers are exactly the same even when grown on the same plant, just as two children born from the same parents are not the same. Even identical twins are not 100% carbon copies of one another. (Clones are the beyond the scope of this work.) No two machined parts are exactly the same. No two measurement systems (no matter how much they cost) are the exact same. Assuming instruments are calibrated and in good operating condition, repeated measurements of the same sample will vary around a value. These measurements will form a characteristic symmetric pattern even in the absence of systematic effects purely due to random experimental error. Because we cannot completely eliminate all variation, we must master quantification of variation. In this chapter, we will look closer at random variation and our propensity to see patterns in random events. Once we quantify random variation in the data we are analyzing, we can leave it alone and stop trying to make unnecessary adjustments to the process until we see variation that is outside of the quantified and characterized random variation. This approach allows us to understand the natural capabilities of our system and make better decisions on engineering tolerances and designs.

7.1 PATTERNS

From where we stand the rain seems random. If we could stand somewhere else, we would see the order in it.

Tony Hillerman
in Coyote Waits, 2009

Humans in general really have trouble with randomness. Experiment on yourself by trying to write a random sequence of 10 numbers. Take a moment to write the numbers down before you read any further. Think about how you arrived at those numbers. Did you alternate between high and low? Did you avoid replicates? Did you avoid replicates next to one another? As Ed Catmul writes in *Creativity, Inc.*, "We can store patterns and conclusions in our heads, but we cannot store randomness itself. Randomness is a concept that defies categorization; by definition, it comes out of nowhere and can't be anticipated" (Catmul 2014). Randomness, by its very nature, has no pattern.

The real problem for us humans is that we see patterns in completely random physical phenomena. Recently on Facebook, a friend posted a photograph of cloud formations in the sky (see Figure 7.1). Her comment below the post was "A dancer." Someone else commented on the photograph, "An angel." On a cloudy Fourth of July afternoon, snuggled up on a picnic blanket, we point out rabbits, cats, horses with chariots all painted with ease on that great canvas, the sky. Initial images of Mars came back from NASA probes; we saw a man walking on the planet or a rock formation that looked like a human face. Photographs showing light interacting

FIGURE 7.1

Random cloud patterns on an afternoon in the Sonoran Desert. What do you see? (Courtesy of Mary M. Walker, used with permission.)

with dust particles are supposedly proof that ghosts exist (Novella 2016). These are examples of pareidolia—seeing a familiar pattern in random data.

We unconsciously see patterns in our daily lives in order to organize our actions and predict responses of the people around us. This is evidenced in our superstitions. The absence of 13th floors in hotels in the United States and full moon phenomena are all examples of superstitious patterns. We consciously find patterns (and assign meaning to them) in clouds, in stars, and all around us, even when they don't really exist.

We also look for patterns in our everyday lives—independent of whether they are true patterns or just random events. We examine these things as if they were patterns and not randomness. When viewing or analyzing data, our *Lazy System 1* will cause us problems with the common logical fallacy of confusing correlation with causation. Recall our Lazy System 1 discussion from Chapter 5 (Kahneman 2011, Novella 2016). Much time and efforts go into "snooping out" fake patterns. The human brain processes and recalls information using pattern recognition. "Science is partly the task of separating those patterns that are real from those that are accidents of random clumpiness. ... The only way to navigate through the sea of patterns is with the systematic methods of logic and testing that collectively are known as science" (Novella 2007). We want to understand our world, make sense of it, and yet we keep running into randomness and the role it plays in our lives, our experiments, and our data.

We make assumptions about the world in an attempt to make it more predictable. We continually try to make the world fit into this model of patterns that we've built from our experiences. We are really good at pattern recognition. The very nature of the human brain function is pattern recognition. "Random information is likely to contain patterns by chance alone," Carl Sagan said. "Randomness is clumpy." Think of rolling two dice. How likely is it that you will roll double sixes three times back to back? It may not happen often, but if you roll dice enough it will happen eventually. Our Lazy System 1 is lousy at recognizing when these patterns are real and when they are not real (Kahneman 2011). As scientists, we are tasked with identifying real and random patterns. "Our 'common sense' often fails to properly guide us, apparently being shaped by evolution to err hugely on the side of accepting whatever patterns we see" (Novella 2016).

Recognition of patterns is not a bad thing. It is natural to observe patterns and correlations. In fact, it is difficult to unsee a pattern or trend that we see. My favorite cartoonist, the creator of *Dilbert*, Scott Adams, actually encourages us to look for, pay attention, and leverage patterns in our lives. He cautions, and I agree, to do this carefully (Adams 2013). Although random variation may initially appear to have a pattern, causation should be established using scientific methods and statistically valid techniques.

Recently, a new field of research and study has developed whose primary objective is pattern hunting (data mining, data analytics, etc.). As digital technologies allow us to tap into old data, we now have unprecedented opportunities to use information collected on paper forms, in files, and in forms. These data open doors for data mining and data analytics with sophisticated specialized algorithms. The application of these data is used in fields from marketing to medicine; even social scientists use pattern hunting in their research (Brown 2010, Dormehl 2014).

7.2 SIMPLE STATISTICS

Variation exists; therefore, we need some simple methods of describing it. We can use a histogram to graphically display the data, but there are times when we'd like to describe the distribution mathematically. In this section, we will cover some of the more common mathematical calculations we use to describe a data set.

We commonly use two statistics to describe the middle or center of our data: the mean and median. The mean is where the middle of the data is; it's a measure of location. With n samples, the sample mean is defined as

$$\mu = \frac{\sum_{i=1}^{n} x_i}{n}. \tag{7.1}$$

Another way to represent the middle of a data set is with the median. The median of a data set is found by first sorting the sample data from smallest to largest; then the median is the middle value, if there is an odd number of data points, or the average of the two middle values if there is an even number of data points.

Let's take a simple example. I roll two dice six times. By summing the two dice from each roll, I get a series of six numbers: 9, 11, 11, 5, 10, and 7. The average or mean is calculated to be $\mu = \frac{53}{6} = 8.83$. In this example, if we sort the numbers from smallest to largest, we get 5, 7, 9, 10, 11, 11. We have six data points (even), and the two middle values are 9 and 10, so the median is $m = 9.5$.

The mean and median are both measures of the center of a data set; different aspects of the data affect them. How are the mean and the median different? The median separates the data into two equal parts. Of course, the value of the data is essential for ranking only, but the values of the value on either side of the median do not affect the median. Each data point is given equal weight or value independent of how extreme it is. In other words, the median is independent of the tail values. The median wouldn't change if our value of 11 was replaced with 24 in the previous example. This is not true for the mean. The mean is sensitive to extreme values. Very different data sets could have the same mean and median but look completely different.

Imagine a horizontal line is a seesaw. If we placed one pound weights on a horizontal line at the values of the data set, and the axis itself had negligible weight, the mean would be the point on the horizontal axis that is in balance. The mean acts as a fulcrum to balance the system of weights. If there were many data points, our seesaw would begin to look like a frequency diagram. A frequency diagram is a histogram with the frequency of occurrence plotted on one axis. An example of a frequency diagram is shown in Figure 7.2.

How can we use this? Let's say we work for a robotics company. We learn that a small titanium rod is a critical piece in the robot project. The assembly is randomly failing and the manager suspects that the rod might be out-of-specification, causing the failure. To determine if this is the case, we've been asked to measure a sample of the next incoming lot of titanium rods. The 30 measurements are listed in Table 7.1. The plotted data give us a frequency distribution (Figure 7.2). The specification for this part from the engineering drawing is $3.2^{+0.50}_{-0.15}$ mm, which means that the window of acceptable lengths is 3.05 to 3.70 mm. We can do some calculations to summarize what we see in the data set. The median is 3.35 mm and the mean is 3.39 mm. It is easy to see that the measurement data easily fits within the dimensional window allowed (3.05 to 3.70 mm) on the drawing. However, neither the mean nor the median is adequate for describing the dispersion of a data set.

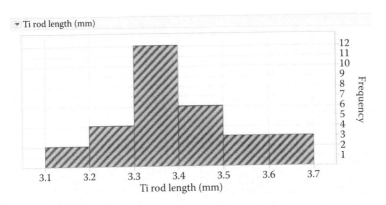

FIGURE 7.2
Frequency diagram of the Ti rod length measurements from Table 7.1.

TABLE 7.1

30 Measurements of Length on the Titanium Part

3.54	3.5	3.46	3.31	3.46	3.15
3.63	3.45	3.33	3.46	3.12	3.31
3.68	3.33	3.35	3.28	3.29	3.26
3.67	3.37	3.33	3.3	3.38	3.45
3.57	3.35	3.23	3.3	3.34	3.44

The easiest dispersion measure that comes to mind might be the range. We can use range to measure dispersion in a data set, which gives us a feel for the spread in the data. The range, R, is simply the difference between the largest and smallest values in the set of numbers or the maximum and minimum values.

$$R = x_{max} - x_{min} \tag{7.2}$$

The range for the data in Table 7.1 and Figure 7.2 is 0.56 mm. A disadvantage of the range is that it completely relies on the extreme data points. The range tells us how far apart the boundaries are, but nothing about what's in between the boundaries.

Let's recap the discussion so far; in this section we've discussed two values that are commonly used to describe the center or middle of a data set and one value that we can calculate to provide information about the boundaries of our dataset. However, neither mean, median, nor range alone nor together gives us enough information about the data to completely represent the whole set of data. We need a different parameter to completely describe a random set of numbers. The mean, a measure of central tendency, will give us the location of the center of our data, but we still need something to accurately describe the dispersion. Recall that the mean, μ, of a set of n values is

$$\mu = \frac{\sum_{i=1}^{n} x_i}{n}. \tag{7.3}$$

Note that μ is also the highest point on the frequency distribution histogram. The data are roughly symmetric about this mean, which is thick with data points at the center and sparser at either end.

Recall in an earlier paragraph that one disadvantage of simply using the range as our measure of dispersion is that it puts so much weight on the extreme values. These extreme values tell us little or nothing about what's happening in the center of the data set. Another disadvantage in using the range to represent the spread in the curve is that from the range we know nothing about how narrow or flat the distribution of data is. We need a different variable to accurately and uniquely identify the curve. We need the standard deviation. The standard deviation is a measure of variation about the mean. The spread in the data can be represented by the standard deviation of the distribution. The standard deviation, σ, describes the deviation around the mean value.

$$\sigma = \sqrt{\dfrac{\sum_{i=1}^{n}(x_i - \mu)^2}{n}} \quad\quad (7.4)$$

With the standard deviation, we are no longer subjected to the effects of extreme values. The standard deviation tells us if the distribution of the data is narrow or wide. The standard deviation tells us exactly how the data are dispersed. The dispersion or width of the curve of measured data will vary depending on many factors. The primary factors that affect the dispersion are type of measurement performed, care used in performing the measurements, and quality of the equipment used to make the measurements. By the way, another advantage of the standard deviation is that we can also say what proportion of our measurements falls within any specified limits, which we will discuss further in the next section.

What have we done so far with simple statistics? We've learned that with the mean and standard deviation for a set of random data set, we can completely describe both where the data are centered and how the data are dispersed. Now let's go a step further.

7.3 IT'S NORMAL

In the presence of randomness, regular patterns can only be mirages.

Daniel Kahneman

Historically, scientists saw random variation in measurement as a failure in the measurement process, not an inherent component of metrology (Mlodinow 2008). After all, if we make a measurement the same way each time, we should theoretically get the same results, right? The early reports of physical measurements yielded a single measurement result. This single reported result was often the measurement (or calculation) the scientist felt was the most careful or the best, unbeknownst to us reading it many years later. These scientists chose some "golden number" to publish from all of their measurements. It wasn't until the turn of the nineteenth century that scientists and mathematicians really engaged in the task of understanding and quantifying random error.

Sir Isaac Newton was one of the first and for many years the only scientist to use the mean value to represent his measurements (Mlodinow 2008). It wasn't until Marquis Pierre-Simon de Laplace, born 120 years after Newton's death, that experimental physics began to become "mathematized." Laplace's work, along with Antoine Lavoisier and Carl Friedrich Coulomb, led to the then new field of mathematical statistics and one of the most important mathematical descriptions of all time, the normal distribution, also known as a bell curve or Gaussian distribution. The initial characterization can be credited to Abraham De Moivre's *The Doctrine of Chances*, in which he describes the bell shape of the curve. The curve is named for Carl Friedrich Gauss, an eighteenth century German mathematician, who demonstrated that repeatedly measuring the same astronomical phenomenon produced a continuous pattern. Gauss was the first to use the pattern that became known as the normal distribution. (His derivation was invalid by his own admission.) Laplace's contribution was connecting the central limit theorem and this continuous normal distribution. Belgian astronomer Adolph Quetelet established the connection between the histogram and the bell curve in 1870, toward the end of Gauss's life. In the 200 years from Newton and De Moivre through the lives of Gauss, Lavoisier, Laplace, and Quetelet, the mathematical descriptions of randomness were developed.

Here's where the beauty of the discussion in the last section begins to become obvious. Recall that given a mean and standard deviation, we can completely describe a set of random data. Given a mean and standard deviation, we can also create the equation for a normal distribution, a Gaussian or bell shaped curve that fits the data set. The mathematical equation for the normal distribution is given by

$$f(x) = \frac{\exp\left(-\frac{1}{2}\left(\frac{x-\mu}{\sigma}\right)^2\right)}{\sqrt{2\pi}\sigma}. \tag{7.5}$$

These two values allow us to draw a bell-shaped curve over the frequency diagram, creating a normal distribution (see Figure 7.3). The ends of the normal distribution are called tails. Note that the mean and median are the same for a normal distribution curve. The mean defines the center or peak of the distribution, while the standard deviation gives the shape of the curve.

The normal distribution is both beautiful and powerful. The normal distribution is a well-defined curve given by Equation 7.5, which is determined

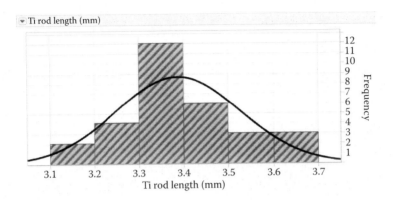

FIGURE 7.3
Frequency diagram fitted with a normal curve of the length measurements from Table 7.1. The normal curve fitted to the data gives a mean of 3.39 mm and a standard deviation of 0.14 mm.

simply by knowing the mean and standard deviation. The curve will be symmetrically drawn around the mean value in a bell shape. The mean, μ, is the middle of the distribution and the measure of the spread (dispersion) in the data is the standard deviation, σ.

Notice in Figure 7.3, the overlayed curve isn't an exact fit to the frequency distribution of the data. With a limited or sample data set, it is unlikely that any curve will fit the frequency diagram exactly. However, in addition to the visual or eye-ball test for fit, there is a more rigorous test of fit of the curve to the data called a goodness-of-fit test. Software packages like JMP® will automatically calculate this for you (JMP 2016). We can also use a normal probability plot and a Tukey outlier box plot to support our use of the normal distribution or the decision not to use it. Also, I should mention that as you delve deeper into the topic of distributions, you will learn that there are many different distributions, and it is important that you select the best distribution for your data. We will limit the discussion in this chapter to the normal distribution.

Now a bit more about the goodness-of-fit testing. Statisticians use something called hypothesis testing to determine goodness of fit. Hypothesis testing is used for many other tests as well, but limit our discussion to the context of goodness of fit. For goodness of fit, the null hypothesis states that the data are from a normal distribution. The goodness-of-fit test will calculate a p value, which can be used to determine whether to reject the null hypothesis or not. Typically, if the p value is small (<0.05), the null hypothesis can be rejected. If you've ever spoken with a statistician, you know that they are very noncommittal. The hypothesis test only allows us to reject the null hypothesis; it says nothing about whether the null hypothesis is actually true. Are you having fun yet? For the data in Table 7.1, the goodness of fit test gives us a p-value of 0.29 which is greater than 0.05. We still don't know if the data are from a normal distribution, but we cannot reject the normal distribution either. More information can be found on *p* values and hypothesis testing in the references provided in Chapter 12.

I mentioned two other indicators that we can use to test whether a data set is from a normal distribution: the normal quantile plot and the Tukey outlier box plot. For the data in Table 7.1, the normal probability plot and the Tukey outlier box plot are show in Figure 7.4. The normal quantile plot allows us to graphically determine whether or not a data set can be approximated by a normal distribution. If the data can be fitted with a diagonal line, this indicates that the normal distribution is good. Departures from

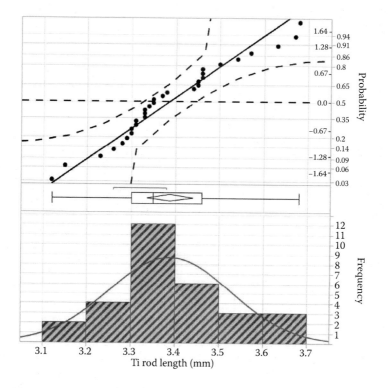

FIGURE 7.4

Frequency diagram fitted with a normal curve of the length measurements from Table 7.1 along with a normal probability plot and a Tukey outlier box plot.

this straight diagonal line indicate departures from normality. The previous normal quantile plot plots the Ti rod length data along the *x* axis and the probability (0 to 1) from something called the cumulative distribution function on the *y* axis. The secondary scale on the *y* axis plots the quantiles from the standard normal distribution, where $\mu = 0$ and $\sigma = 1$. Quantiles simply split the data into bins based on percentages, where the median is the 50th percentile and the 25th and 75th percentiles are called the quartiles.

The other graph in Figure 7.4 is the Tukey outlier box plot, which is useful in identifying potential outliers. If an outlier exists in the data set, it will be highlighted in the Tukey outlier box plot (Tukey 1977). The Tukey outlier box plot divides the data into four groups. The box contains 50% of the data. Each end of the box has whiskers, which extend from the box, which show any mild outliers. Any data outside the whiskers are considered an extreme outlier. In this case, there were no outliers highlighted by the Tukey outlier box plot. I will leave further explanation about this graph and Tukey outlier box plot to further research (see Chapter 12) and the statistical software package you choose to use.

As I alluded to earlier in the chapter, the standard deviation divides the normal curve into equal length multiples of the standard deviation about the mean as shown in Figure 7.5. We see that one standard deviation on either side of the mean represents 68.27% of the population (Figure 7.5a). In other words, we expect that 68% of the population will be within one standard deviation on either side of the mean. The ordinates erected at one standard deviation on either side of the mean include 68.27% of the area under the curve. Two standard deviations on the either side of the mean represent 95.45% of the population (Figure 7.5b). Likewise, three standard deviations on either side of the mean represent 99.73% of the population of the random data set.

This is what's so beautiful about the normal distribution. There are many normal curves, but they all share the same characteristic density properties described with the 68%–95%–99.7 rule, which is also called the *Empirical Rule*. One standard deviation on either side of the mean contains 68% of the data (from $\mu - 1\sigma$ to $\mu + 1\sigma$). Two standard deviations on either side of the mean contain 95% of the data (from $\mu - 2\sigma$ to $\mu + 2\sigma$). Three standard deviations on either side of the mean contain 99.7% of the data (from $\mu - 3\sigma$ to $\mu + 3\sigma$). For a normal distribution, almost all the data are contained within three standard deviations of the mean (Figure 7.5c) and the complete area under the normal curve represents 100% of the population.

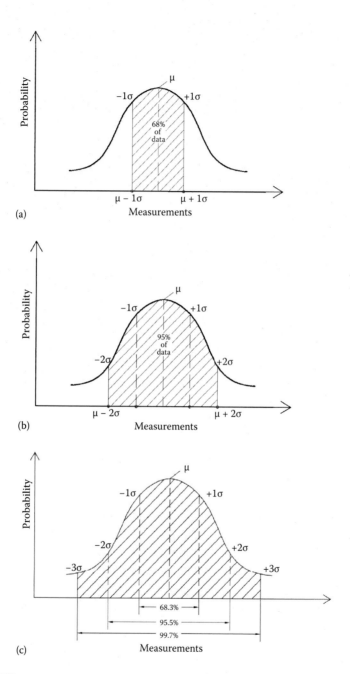

FIGURE 7.5

Normal distribution curve showing the data within (a) one standard deviation of the mean accounts for 68.27% of the population, (b) two standard deviations of the mean accounts for 95.45% of the population, and (c) three standard deviations of the mean accounts for 99.73% of the population.

7.4 IT'S NORMAL, SO WHAT?

What can we learn from the normal distribution curve? Unless we can collect all the data (or an extremely large sample data set), when we plot our data, we will get a frequency distribution histogram. However, we can calculate from our data the curve that would result should we have the whole population of data to plot. I say this because our histogram may not look exactly like the curve generated from the calculation and plot overlay, as we saw in Figure 7.3.

The news is bombarded with studies that use samples and make conclusions about a larger population. You may have wondered, how is this possible? What assurances do we have that a sample represents the population from which it originates? The credit for allowing us to make inferences from samples as representative of a population goes to the central limit theorem. The foundation of the central limit theorem is that a sample will resemble the population from which it is taken (that is, assuming that the sample size is sufficiently large and that it was properly taken. Assume this every time you see an asterisk next to a sample). The central limit theorem is a powerful conclusion of modern statistics that allows us to make inferences (Wheelan 2013).

1. With information about a population, we can infer things about a sample*.
2. With information (μ, σ) about a sample*, we can infer things about the population from which the sample was taken.
3. With data on a sample and a population, we can infer whether or not the sample is likely to have come from the population.
4. If we have information about two samples, we can infer whether both samples originated from the same population.

The central limit theorem tells us that a frequency distribution diagram of the sample means of any population can be fit with a normal distribution that centers around the population mean. What does this imply? Let's say instead of 30 samples of the Ti rod one time, we repeat the process 30 more times, each time throwing the 30 measured parts back into the box of 10,000 parts. The new sample measurement mean is calculated from a new random sample of 30 parts each time. The central limit theorem tells us that the sample means will be close to the population mean and these means will be distributed normally around the population mean. This is true regardless of the shape of the distribution of the whole population. Restated, the average and standard deviation of some characteristics or properties calculated from our sample will not deviate too much from the average of the population. Further, the average and standard deviation of some characteristics or properties calculated from the population will look like a sample drawn from that population. It is important to remember that these implications of the central limit theorem hold true independent of the distribution of the original population.

So according to the central limit theorem, the information we know about our sample will be representative of the whole population. Stated another way, the central limit theorem tells us that a large-enough sample will look like the population from which the sample originates. The larger the sampling size of each sample population, the better this approximation is, giving a more accurate mean with a smaller standard deviation. The question to ask then is "What is meant by large?" Obviously, larger sample sizes will come closer to normal distributions than smaller sample sizes. The larger the sample size, the less effect one data point will have on the overall distribution, but our sample size needs to be at least 30 for the central limit theorem to hold true. Although repetition alone, 30, 60, or 100 times, doesn't ensure repeatable and reproducible measurements, it does help us begin to isolate sources of variation (Youden 1962).

Another side note comment on representative data, when statisticians talk about mean and standard deviation for a population and sample, they tend to change the notation from μ and σ to $\bar{x}$ and s, respectively. The value of n now represents the sample size rather than the population size. The sample mean and standard deviation are now

$$\bar{x} = \frac{\sum_{i=1}^{n} x_i}{n} \tag{7.6}$$

and

$$s = \sqrt{\frac{\sum_{i=1}^{n} (x_i - \bar{x})^2}{n-1}}. \tag{7.7}$$

The mean and standard deviation are now the summary statistics that represent the sample that represents the population. Notice that the denominator for the standard deviation has changed for the sample. Using "$n - 1$" in the denominator compensates for underestimating the dispersion or spread in the population. Because "$n - 1$" is just a bit smaller than "n," this will make the standard deviation larger for the sample.

Now, let's look at how to determine the dispersion in the sample means. In other words, we want to know how closely all the sample mean values cluster around the population mean. To answer this inquiry, we need to look at the standard error, *SE*. The standard error is defined as

$$SE = \frac{s}{\sqrt{n-1}}, \tag{7.8}$$

where n is the sample size and s is the standard deviation. It is important to not get standard deviation and standard error mixed up. Standard deviation measures the dispersion in the population, while standard error measures the dispersion in the sample means. Notice that they are related. The standard error depends on both the standard deviation of the sample and the sample size. A large standard error indicates that we have a large standard deviation or a small sample size. In other words, the sample means are not clustered but are potentially highly spread out around the population mean (Wheelan 2013).

Now, going back to our powerful central limit theorem, we know that the sample means are normally distributed. This tells us that 68.2% of the means lie within 1 standard error of the population mean, 95.4% lie within 2 standard errors, and 99.7% lie within 3 standard errors.

We might also be interested in relative variability, where the most common measure is the coefficient of variation, which is simply the ratio of the standard deviation to the mean.

$$CV = \frac{s}{\bar{x}} \tag{7.9}$$

We often use the coefficient of variation to represent the process non-uniformity (Wortman et al. 2007). For example, when we measure the etch depth on a silicon wafer at 49 locations on a wafer, we can use the coefficient of variation as a measure of the relative variability of the etch process on that one silicon wafer.

Some people see random, unforeseen events as something to fear. I am not one of those people. To my mind, randomness is not just inevitable; it is a part of the beauty of life.

Ed Catmul

As a simple example of normality concept, let's take a simple paper clip breaking experiment. Professor Henry Petroski has used this in his engineering classes at the University of Virginia (Petroski 1982). Open the paper clip flat, then bend the clip back and forth. Count how many bends until the clip breaks. Repeat the "bend to break" activity for a large number of paper clips, roughly 30. The more data points gathered, the better. Record the numbers of bends and plot the data on a frequency diagram. The results may fall into a bell-shaped curve. The more data gathered, the more well defined the curve will become. In this example, we are demonstrating that all of the clips aren't equally strong (assuming that the clips are all bent using the same bending technique and strength). If you repeat this with a friend or classmate, not only will the subtleties of the bend technique be different but the different strengths will also come into play. A plot of the data should show a normal distribution, but it may have a large standard deviation. My son decided we'd try this. Figure 7.6 is the data set of my son's (Ben) paper clips breaks. Notice that the frequency diagram overlayed with a beautiful normal probability curve defined by the mean number of bends is 5.9 with a standard deviation of 1.2.

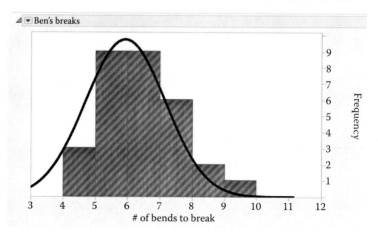

FIGURE 7.6

Frequency diagram of how many bends it took for Ben's paper clips to fail (break). The data have been fit with a normal curve defined by a mean number of bends of 5.9 with a standard deviation of 1.2.

7.5 DARK SIDE OF THE MEAN

Once we've collected our data sample and described the random variation with a Gaussian or normal distribution function (i.e., calculated $\bar{x}$ and s), we can learn a few things by taking a closer look at the distribution. The first thing we can do is to compare μ to the *true value*, if we have some approximation for the *true value*. How close are these two numbers? The closer μ is to the *true value*, the better approximation it is for the *true value*. The more the measurements cluster around the *true value* and the smaller the value of s, the better approximation μ is to the *true value*. If the measurements do not cluster around the *true value*, it is possible that μ is not a good approximation for the *true value*.

A primary attribute of the normal distribution curve is that small deviations occur more frequently than large ones. When we begin to see large deviations resulting in a large cluster of data far away from the mean, this typically means that it's time to look for a nonrandom cause. As an example of this, I repeated the paper clip experiment that Ben did in the last section. The results are shown in Figure 7.7. The data are fit with a normal distribution curve defined by a mean number of bends for my breaks of 6.5 with a standard deviation of 2.4. Notice that the mean number of bends and the standard deviation are both higher than Ben's. The higher mean and standard deviation are being driven completely by the three outliers that the Tukey outlier box plot identified, the large numbers at the

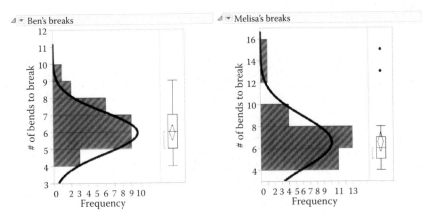

FIGURE 7.7

Frequency diagram showing Ben's and Melisa's bending 30 paper clips to failure. The data are fit with a normal curve defined by a mean number of bends for Melisa's breaks of 6.5 with a standard deviation of 2.4.

top of the tail. This was interesting. Did I happen to grab three paper clips that were twice as strong as the others in the container or was something else going on? Should we assume that the manufacturer of the paper clips mixed materials or made a few of the paper clips thicker? When there are outliers in a data set, it is important to try and identify root cause or what might have caused it. When I reviewed the data in the order in which I broke the paper clips, I noticed that these outliers were my first three attempts at bending to breaking of the paper clips.

It could be that I didn't have my bending technique perfected initially. Therefore, because I had plenty of paper clips, I just repeated the whole experiment. I was now an experienced paper clip breaker. The second set of break attempts can be found in Figure 7.8. The data are fit with a normal curve defined by a mean number of bends for my second attempt of 5.9 and the standard deviation of 1.4. Notice that the distribution in Figure 7.8 has the same mean as Ben's experimental results of 5.9, but Ben still has a slightly smaller standard deviation. This simple example is perfect for illustrating that nonrandom sources can lead to false conclusions or skew a data set. It is never enough to only examine a data set in one way and draw conclusions. If possible, plot the data multiple ways, looking specifically for nonrandom sources in the data.

The best way to address outliers is to repeat the experiment. With enough repeats, the outliers in the tail of the distribution have less effect on the mean and standard deviation of the distribution. Technically, although we reran the experiment, we wouldn't want to throw out the data

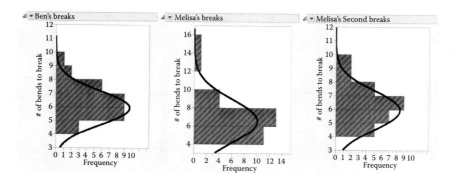

FIGURE 7.8

Frequency distribution of Melisa's second round of bending 30 paper clips to failure. The data are fit with a normal curve defined by a mean number of bends for Melisa's second attempt of 5.9 and a standard deviation of 1.4.

from the first experiment and only use the second round, unless we knew that something had gone wrong. In this example, we see that it's simply my lack of practice and inconsistent method that caused the variation. Therefore, I may want to include both experimental trials worth of data. This new combined frequency diagram and normal distribution curve are shown in Figure 7.9. Notice that the outliers still have an effect on the results even with double the number of data points. The data are fit with a normal curve, which is defined by a mean of 6.2 and a standard deviation of 1.96 for my combined attempts. The mean and standard deviation for the combination of the trials is in between the results for the individual first ($s = 2.4$) and second ($s = 1.4$) experimental trials.

There are times when we cannot repeat experiments. If something was obviously wrong, for example, recording the wrong units or missing a decimal place, these measurements should be discarded. We should never include data that we know are wrong in our analysis. However, there are times when outliers exist and there is no explainable or obvious reason to exclude them. One statistically based method for "throwing away" unreasonable data points is named Chauvenet's criterion. Chauvenet's criterion provides an acceptable inclusion range about the mean for a data set. Stated another way, Chauvenet's criterion specifies the location on the tail of the distribution beyond which we can reject those data points. The criterion specifies that any point may be rejected if the probability of

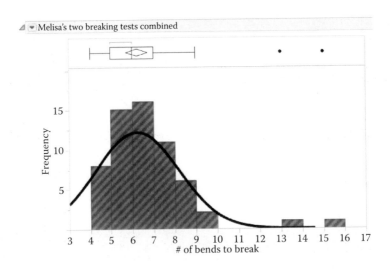

FIGURE 7.9
Frequency distribution of Melisa's first and second round of bending paper clips to failure.

obtaining the point is less than ½n, where n is the number of data points. "If the expected number of measurements is at least as bad as the suspect measurement is less than ½, then the suspect measurement should be rejected" (Lin and Sherman 2007).

There are multiple methods for implementing Chauvenet's criterion (Coleman and Steele 1999, Kirkup 2002, Lin and Sherman 2007, Taylor 1982). One of the simplest is presented by Lin and Sherman (2007) and doesn't involve looking points up on normal distribution tables. Assume we have a dataset and would like to identify any outliers. For all n data points, calculate the mean and standard deviation. Use the following expression to reject a suspicious data point, x_i:

$$n \times erfc\left(\frac{|x_i - \bar{x}|}{s} \right) < \frac{1}{2}. \tag{7.10}$$

The function $erfc(x)$ is the complimentary error function. This function can be calculated easily using online calculators. After using Chauvenet's criterion, if a point is thrown out, we will have a completely new distribution with a different number of data points, n', a new mean, $\bar{x}'$, and a new standard deviation, s'. We now have two distributions, one with an outlier and one without the outlier. It's important to make every effort to understand what happened and why the data set contained outliers to begin with. An example lab report from Alex Cress and Briana Fees demonstrating this technique is shown in Figure 7.10. They participated in an experiment to measure the hardness of stainless steel discs using a Rockwell Hardness Tester in Material Engineering 210 at San Jose State University.

Now for a word of caution, this process is somewhat controversial among some scientists and engineers who do not approve of its use. Dr. John Talyor's book, *Introduction to Error Analysis*, has a great discussion of this use of Chauvenet's criterion and the controversy surrounding its use (Taylor 1982). For practical purposes, I think we need some test to account data that are unreasonable, but as scientists and engineers, you will need to develop your own stand on this. I would add that anytime Chauvenet's criterion or other similar techniques are used, it should be mentioned so that everyone who reads or hears your findings, results, and/or conclusions will be aware. Throwing away data points is not something to be done without incontrovertible explanations. As computational output has become so easy in recent years, building in an error checker is a

Hardness Study of 304 Stainless Steel Discs

Alex Cress and Briana Fees
Material Science Engineering 210
San Jose State University

1.0 Introduction

Random variation is inherent in any process or measurement. With a large enough sample of repeated measurements, if the variation is purely random in nature, the data can be fit with a normal distribution. The normal distribution is defined by two quantities, the mean of the distribution and the standard deviation in the data. By controlling all critical inputs and minimizing other process variation, the random variation can be estimated with the mean and standard deviation. In this experiment, the hardness of 30 discs machined from 304L stainless steel rod were measured. The lab objective was to quantify the random variation of the hardness measurement process.

2.0 Material

The Carpenter Technology Corporation Certificate of Conformance for stainless steel rod showed a hardness of 84 HRB with the elemental weight percentages shown in Table 1, with the balance being iron.

A rod of 304L low carbon stainless steel was machined into discs.

Table 1: Elemental weight percentage of the discs.

Element	C	Mn	Si	P	S	Cr	Ni	Mo	Cu	Co	N
Weight%	0.018	1.72	0.58	0.032	0.024	18.28	8.37	0.38	0.33	0.34	0.069

The machining process used ATTAR-C® lubricant. Samples were cleaned prior to measurement to remove any cutting fluid residue. A total of 30 discs were provided for measurement. Stainless steel discs had a diameter of 1 1/2 inches and thickness of 1/2 inch.

FIGURE 7.10

Sample lab report from Alex Cress and Briana Fees demonstrating the use of Chauvenet's criterion. *(Continued)*

3.0 <u>Measurement</u>

Hardness measurements were performed in the HRB scale using an uncalibrated Wilson 500 series Rockwell Hardness tester and a 1/16" ball tip. The samples were split into 6 groups of 5 samples at random, with each measurement performed one time on all 30 samples. The measurements purposefully targeted areas away from any edge machining effects.

4.0 <u>Method</u>

An Input-Process-Output diagram, Figure 1 was created to help confirm that the only input variable we were deliberately changing was the stainless steel discs. Input items are labeled with C for controlled, N for noise or uncontrolled and X for intentional variation.

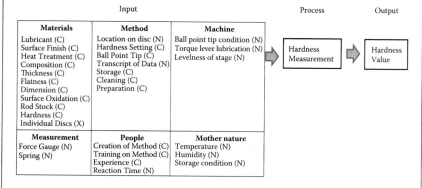

Figure 1: Input-Process-Output Diagram for obtaining hardness measurements.

The normal quantile plot can be used as a test for normalcy. Here, if the data is normally distributed, the normal probability plot will be represented by a linear, diagonal line. This allows for a visual evaluation of how well a data set is normally distributed, and helps identify possible outliers in the data set. The Tukey outlier box plot can be used to further identify outliers in the data. We can quickly visualize the 1st and 3rd quartiles bounding the central 50% of data and as a larger probability range defined by 'whiskers', as well as how the data falls within these ranges. When data falls outside of this

FIGURE 7.10 (CONTINUED)
Sample lab report from Alex Cress and Briana Fees demonstrating the use of Chauvenet's criterion.
(Continued)

defined range, it is often termed an 'outlier'. The data visualization and analysis were performed using JMP® statistical software.

5.0 Results and Discussion

The hardness measurements are given below in Table 2. The maximum and minimum values are 89.1 and 79.8, respectively.

Table 2: Hardness measurements of each disc.

A	86.8	C	85.6	E	84.8
A	86.6	C	87.3	E	84.3
A	86.4	C	85.2	E	86.3
A	85.9	C	88	E	85.1
A	89.1	C	86.2	E	86.3
B	86.4	D	86.6	F	87.7
B	84.6	D	84.1	F	85.8
B	84.1	D	85.3	F	84.2
B	85.1	D	79.8	F	85.3
B	85.7	D	85.3	F	88.6

Figure 2 shows the distribution of the measurement data overlaid with a fitted normal curve. The Tukey outlier box plot is shown on top with two outliers highlighted, the minimum and maximum values. The normal quantile plot indicates one data point that may be of concern.

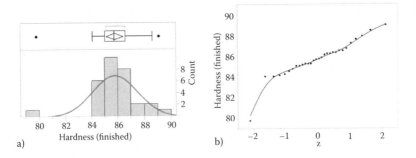

Figure 2: a) Distribution of measurement data and Tukey outlier box plot of hardness measurement data, showing a mean hardness of 85.8 with a standard deviation of 1.7, b) normal quantile plot of hardness data.

FIGURE 7.10 (CONTINUED)
Sample lab report from Alex Cress and Briana Fees demonstrating the use of Chauvenet's criterion.   (*Continued*)

Both the histogram and the outlier box plot clearly show a data point that is disconnected from the distribution, a suspected outlier. Chauvenet's criterion is a test that can be used for the determination of whether data that is shown to be significantly distant from the mean is 'ridiculously improbable'. By applying Chauvenet's criterion, the outlier can be evaluated to determine 'reasonableness'.

Assume that we have made N measurements of quantity x. In our case, we have made 30 measurements of hardness. The mean is $\bar{x} = 85.8$ and standard deviation $\sigma = 1.7$. Both from visual observation of the distribution and using the outlier box plot, the data point $x = 79.8$ is a suspicious measurement. In other words, 79.8 looks different from the rest of the population and is far away from the mean. The quantity t, the number of standard deviations which our suspect data point differs from the mean, is defined mathematically in Equation 1.

$$t = \frac{(\bar{x} - n)}{\sigma} \qquad \text{Equation 1}$$

In this case, the calculated value for t is 3.5, indicating that the suspect data point is 3.5 standard deviations away from the mean. Next, we want to calculate the probability that the suspect point is outside the main distribution by at least 2 standard deviations. p_o is given by Equation 2, and the error function, erf, is operating on t.

$$p_o = 1 - erf(t/\sqrt{2}) \qquad \text{Equation 2}$$

The probability, P, is calculated by taking the product of the number of measurements, N, and the probability outside, p_o.

$$P = Np_o \qquad \text{Equation 3}$$

Applying these equations to our data set for the outlier as $n = 79.8$, we find a value for $P = 0.0143$, which is well below the traditional cutoff of 0.5. Using Chauvenet's criterion, we can say that the data measurement of $n = 79.8$ doesn't meet our 'reasonableness' criteria.

FIGURE 7.10 (CONTINUED)

Sample lab report from Alex Cress and Briana Fees demonstrating the use of Chauvenet's criterion.

(Continued)

After removing this measurement from the data set, the data can be replotted. By repeating the process once again of fitting the data to a normal curve, the new distribution can be seen in Figure 3. As expected, the mean was only slightly affected by the outlier but the standard deviation was strongly affected. The mean hardness measurement increased from 85.8 to 86.0, while the standard deviation dropped from 1.7 to 1.3.

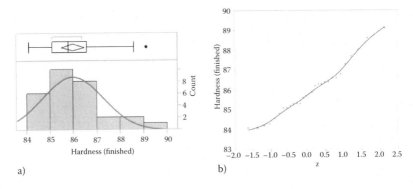

a)   b)

Figure 3: a) Distribution and Tukey outlier box plot of modified hardness data, showing a mean of 86.0 and a standard deviation of 1.3, b) normal quantile plot of hardness data with the outlier removed.

Although the data set now looks better with a tighter distribution, the outlier box plot has identified another potential outlier at 89.1. While repeated use of Chavenets' criterion is discouraged by many scientists, performing the same calculations with this outlier against the original data set gives some interesting results. Using the value of $n = 89.1$, we find a probability of 1.57, which is > 0.5, thereby, prohibiting a reasonable exclusion.

6.0 Conclusions

The random variation of hardness measurements on 30 samples of a machined 304L stainless steel rod was measured. The mean hardness was determined to be 86.0 with a standard deviation of 1.3. The mean hardness measurement is in reasonable agreement with the value provided by Carpenter (84.0). The standard deviation magnitude is indicative of random variation or noise in the measurement inputs. Although

FIGURE 7.10 (CONTINUED)
Sample lab report from Alex Cress and Briana Fees demonstrating the use of Chauvenet's criterion.   (*Continued*)

outliers were identified in the measurement data using the Tukey outlier box plot, the application of Chauvenet's criterion showed the farthest data point from the mean was more than 2 standard deviations away. Chauvenet's criterion is not 'proof' that this data point was an outlier. The data point was outside 2 standard deviations, and no reasons were found for why this sample would be very different from the others from this rod. We removed the point from the data and recalculated our results. The hardness measurements for the samples ultimately fit a normal distribution, with no indication of any inputs which would lead to systematic variation in the hardness value data. With the random variation characterized for the hardness measurements, future experiments can now be performed on these samples allowing us to be more confident in any changes to hardness due to additional processing.

FIGURE 7.10 (CONTINUED)
Sample lab report from Alex Cress and Briana Fees demonstrating the use of Chauvenet's criterion.

prudent and reasonable approach. For example, for his PhD thesis, David R. Wagner used a combination of analytical techniques combined with error checking to ensure the best results possible (Wagner 2013). Also, we should use Chauvenet's criterion only on a normal distribution; it doesn't work well on multimodal distributions. With multiple modes in the distribution, it may be the case that we have different data sources that need to be better understood before analysis (Lin and Sherman 2007).

Throwing away data, especially outliers, could be a big mistake. The outliers could be the most interesting part of a data set. These are often points worthy of investigation in order to understand why they differ. It is the outliers that might lead to significant discoveries. One such example was described in a *Scientific American* article (Benedick 1992). As you may know, satellites measure the ozone level over Antarctica regularly. In the early 1980s, a significant seasonal drop in ozone levels over Antarctica was detected. Scientists analyzing the data subsequently spent two years rechecking their satellite data. The scientists discovered that satellites had been measuring the data and recording a drop in ozone levels over time. However, the program used to analyze the data was programmed to reject outliers and treat it as anomalies. If the computer had been programmed to highlight the outliers, the scientists could have investigated the outliers on the first occurrence. A lot time and resources were wasted by "throwing away" the outliers.

Since we are prone to see patterns in random distributions, we have to take care in assuming that all bell-shaped curves are due to random variation. The central limit theorem does not apply to all systems. When the majority of values fall in the middle and tail off in either direction, we suspect that it is a normal distribution; however, there is another step we should take to confirm. Plotting the data versus time or in order (a control chart) is one way to check. Looking at the frequency distribution in Figure 7.11, we see what looks like normally distributed data. However, if we look at the control chart (Figure 7.12) of the data in the order it was collected, we see that there is a systematic drift over time. Caution should be used at all times when dealing with data (Deming 1982, Flaig 2016, Khorasani 2016).

Another potential issue we may encounter in random distributions occurs when the mean of a data set lies too far to the right or left of the median, we say that it is skewed to the right. *Skewness* is typically measured using Pearson's coefficient of skewness:

$$SK = \frac{3*(\text{mean} - \text{median})}{\text{standard deviation}}. \tag{7.11}$$

In the case of a perfectly symmetrical distribution, mean = median; therefore, $SK = 0$. The skew is important as it may be indicative of a nonrandom process. When we see a skew in our data distribution, it is time to look for a nonrandom culprit. Once a nonrandom source has been identified, it's best to repeat any measurements to confirm that the source has been eliminated.

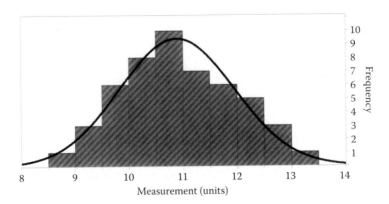

FIGURE 7.11

Frequency histogram of measurements made over 50 days fit with a normal distribution curve with a mean of 10.9 and a standard deviation of 1.06.

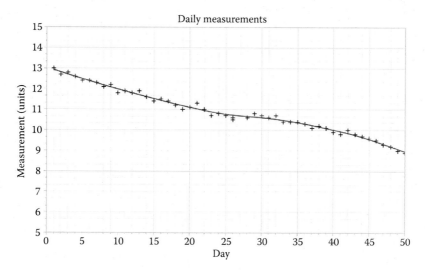

FIGURE 7.12

Normal distribution of a measurement made repeatedly over time. Notice that the distribution of the data is normal but using a scatter plot measurement over time shows a systematic variation.

There are a number of physical processes that are inherently random such as entropy, an important concept in the second and third laws of thermodynamics. The random motion of molecules in gases, liquids, and plasmas exert small forces on the surfaces they encounter. Radioactive decay is another random process. It is impossible to predict when nuclei will emit a particle. These particular fields of study have statistical studies developed to address their particular phenomenon.

7.6 KEY TAKEAWAYS

We humans see patterns all around us. In many cases, it is good to pay attention to patterns on a personal level, e.g., foods that make us feel sleepy or energize us, etc. (Adams 2013). However, not all patterns are real; we want to pay attention to them, but be cautious. When it comes to recognizing patterns, it is essential that we not let clumps of data convince us that anything other than a normal (as in Gaussian) pattern exists when it doesn't. As scientists and engineers, we must use our critical thinking skills with all data. Randomness shows up in all measurements. Randomness is expected. It's normal and that's normal. With our experimental results, it is important that we characterize and quantify the *random variations*. Randomness can be characterized with experimental duplication. The more replicates, the better. However, when there are at least 30 randomly selected sample means from a population, the power of the central limit theorem is on our side. Finally, normal distributions can be used to describe a set of random data provided that there is no systematic variation in the measurement. It is important to explore the data set using various analysis techniques in order to root out any nonrandom sources. Happy experimenting! May the central limit theorem be with you.

P.S. We have reached another milestone in the book. Let's take a moment to reflect on the big picture. We have a problem to solve. We are attempting to accurately measure an experimental response; therefore, we want to control as much of the variation as possible to minimize any uncertainty in our experimental findings. We've learned three ways to deal with variation: (1) eliminate or minimize *unintentional variation* through checklists or *standard operating procedures*, (2) thoroughly characterize our measurement system to quantify *systematic variation*, and (3) quantify any random variation. Now, in Chapters 8 and 9, we will discuss options to exploit certain types of variation (intentional variation) in our experiments in order to explore certain effects.

REFERENCES

Adams, S. 2013. *How to Fail at Almost Everything and Still Win Big: Kind of the Story of My Life*. New York: Portfolio/Penguin.

Benedick, R. 1992. Essay: A Case of Déjà vu. *Scientific American* 266:160.

Brown, B. 2010. *The Gifts of Imperfection: Your Guide to a Wholehearted Life*. Center City, MN: Hazelden Publishing.

Catmul, E. 2014. *Creativity, Inc.: Understanding the Unseen Forces That Stand in the Way of True Inspiration*. New York: Random House.

Coleman, H. W. and W. G. Steele, Jr. 1999. *Experimentation and Uncertainty Analysis for Engineers*. 2nd Ed. New York: John Wiley & Sons.

Deming, W. E. 1982. *Out of the Crisis*. Cambridge, MA: Massachusetts Institute of Technology.

Dormehl, L. 2014. *The Formula: How Algorithms Solve Our Problems...and Create More*. New York: Penguin Group.

Flaig, J. J. 2016. A Bell Shaped Distribution Does NOT Imply Only Common Cause Variation. *The Quality Technology Corner*. www.d577289.u36.websitesource.net /articles/BellCurveNotRandom.htm.

Hillerman, T. 2009. *Coyote Waits*. New York: Harper.

JMP. 2016. http://www.jmp.com.

Kahneman, D. 2011. *Thinking, Fast and Slow*. New York: Farrar, Straus and Giroux.

Khorasani, F. 2016. Private communication.

Kirkup, L. 2002. *Data Analysis with Excel©: An Introduction for Physical Scientists.* Cambridge: Cambridge University Press.

Lin, L. and P. D. Sherman. 2007. Cleaning Data the Chauvenet Way. Paper presented at Southeast SAS Users Group, Hilton Head, SC.

Mlodinow, L. 2008. *The Drunkard's Walk: How Randomness Rules Our Lives*. New York: Pantheon Books.

Novella, S. 2007. Data Mining—Adventures in Pattern Recognition. Neurologica Blog. 2.26.07.

Novella, S. 2016. Your Deceptive Mind: A Scientific Guide to Critical Thinking. Lecture Notes. The Great Courses.

Petroski, H. 1982. *To Engineer Is Human: The Role of Failure in Successful Design*. New York: Vintage Books/Random House.

Taylor, J. R. 1982. *An Introduction to Error Analysis: The Study of Uncertainties in Physical Measurements*. 2nd Ed. Sausalito, CA: University Science Books.

Tukey, J. W. 1977. *Exploratory Data Analysis*. New York: Addison-Wesley.

Wagner, D. R. 2013. Coal Conversion Experimental Methods for Validation of Pressurized Entrained-Flow Gasifier Simulation. PhD diss. University of Utah.

Wheelan, C. 2013. *Naked Statistics: Stripping the Dread from the Data*. New York: W. W. Norton.

Wortman, B., W. Richardson, G. Gee, M. Williams, T. Pearson, F. Bensley, J. Patel, J. DeSimone, and D. Carlson. 2007. *The Certified Six Sigma Black Belt Primer*. West Terre Haute, IN: The Quality Council of Indiana.

Youden, W. J. 1962. *Experimentation and Measurement*. Washington, DC: National Science Teachers Association. The book was reprinted in 1994 by the National Institute of Standards and Technology (NIST) and in 1998 by Dover Publications.

8

Experimenting 101

Out of the library, into the laboratory.

Rallying cry of the first members of the Royal Society

We now have all the pieces of the puzzle to create a reliable, repeatable experiment. We know that we need to control our experimental setup and procedures. We know that we must have a well-characterized measurement system in which we have quantified repeatability and reproducibility. We know that random variation will play a role in our results and how to quantify its contribution. Now, we can confidently begin to explore and experiment by intentionally manipulating variables. The first seven chapters of this book were just a setup to give us confidence as experimental problem solvers.

Although there are a number of different mathematical techniques that can be used, we'll stick with one in this chapter. We'll look at one-factor-at-a-time experimentation and use regression analysis to build a model of the experimental process space. This is a great starting point for any experimentalist. We change one factor and record the effects. We did this in high school and college labs most likely without realizing it. Any time we've fit our data with a line and displayed the equation for that line, we've built a model of the experimental process space. Most of the time, we use some form of regression analysis to create that line. Although many graphic software programs now make this very easy, there is sophisticated mathematics behind the development. It is important that we understand where it all comes from and why we get the results that we get.

8.1 TORTURING NATURE

Recent history (okay, the last few hundred years of history) is rich with scientists eager to learn the "eternal laws that govern the universe" (Dolnick 2011). This history reveals an evolution in the methods that scientists have used to gain this knowledge. Our history begins with detailed observations and evolves to measurement of those observations. Galileo introduced us to indirect measurement techniques in the sixteenth and seventeenth centuries. Sir Francis Bacon brought us empiricism and the scientific method. Nature must be "put to the torture," Bacon declared. By the mid-seventeenth century, the dozen founding members of the Royal Society were calling for experimentation, creating artificial situations and recording observations. Experiments were something new. To the seventeenth century world, this was a radical call. The universities at the time saw it as their responsibility "not to discover the new but to transmit a heritage," according to historian Daniel Boorstin. Students who didn't abide by this philosophy could be fined five shillings (Boorstin 1983). Curiosity, according to Augustine, was the equivalent of lust. The men of the Royal Society wanted to probe, poke, and test, not passively observe the world from behind a curtain (Dolnick 2011).

Our curiosities lead us toward a deeper understanding of the world inside us, around us, and beyond us. Although it may seem that the Royal Society and Isaac Newton existed long ago in an entirely different era, we are still, hundreds of years later, trying to create, optimize, and teach repeatable, reproducible experimental practices.

8.2 PROCESSING, A DEEPER LOOK

In Chapter 1, we learned that to "experiment" meant a test undertaken to make an improvement in a process or to learn previously unknown information. Embedded in this definition is the word *process*. What is a *process*? In order to define a designed experiment, we need to have a good understanding of this term. A process is any activity based on some combination of inputs (factors), such as people, material, equipment, policies, procedures, methods, and environment, which are used together to generate outputs (responses) related to performing a service, producing a product, or completing a task. (Recall from Chapter 3 the Input–Process–Output diagram.)

Process inputs can be any of the following: people, material, equipment, policies, procedures, or methods. Whether an input is controlled or uncontrolled, it can (and will) affect the experimental results. These inputs blend together to create corresponding outputs in the soup pot called process. The outputs are the responses we get after the blending has taken place, after the "process" has acted upon the combination of inputs. As engineers and scientists, our jobs are filled with different processes both inside and outside of the laboratory. Some processes are obvious, such as an industrial braze or anneal process, an injection molding process, a plating, polishing, etch, or deposition process. Mixture development such as optimization and maintenance of asphalt or concrete; creating household items such as bleach, shampoo, and perfume; and even optimizing a cookie recipe are all processes.

Development, optimization, and maintenance of these processes often fall under the responsibility of a development engineer, process engineer, or manufacturing engineer. There are other processes that involve dimensional design of parts used in automobiles, aircraft, space shuttles, or cell phones. Dimensional designs are often developed and established with constraints from both upstream suppliers and downstream customers. Most scientists or engineers will be involved in development, optimization, and maintenance of some process as an important part of their job. Therefore, familiarity with process development, optimization, and maintenance as a part of a formalized, strategically designed problem solving plan is a very nice skill to bring to a new position and an essential skill to hone early on.

The outcome of a well-designed, well-executed experiment is a mathematical model that provides a relationship between the inputs and outputs. In other words, the resulting model is a mathematical description of the process. The mathematical model will contain information on how to optimize and improve the process, how to perform a sensitivity analysis which can be used for tolerance evaluations, and how to reduce variation and possibly make our response robust (insensitive) to factors out of our control.

In conducting a strategically *designed experiment*, we will purposefully make changes to the inputs (or factors) in order to observe corresponding changes in the outputs (or responses). The key word is *designed*. A list of synonyms for design turns up a critical key word: *plan*. Planning implies deliberation and intentionality, which will be covered more in Chapters 9 and 10. The critical point here is that a designed experiment is well planned with thoughtful, deliberately controlled inputs. Going into an experiment with a detailed plan allows us to minimize unexpected events that may make our experiment unreproducible or unrepeatable. Let's look at the simplest design, one-factor-at-a-time.

8.3 THE SIMPLEST EXPERIMENTAL MODEL

With one-factor-at-a-time experimentation, we typically think of varying one factor multiple times in order to better understand the effect on some response variable. The simplest relationship we can hope for is linear. One of the mathematical formulas we learned in our early algebra classes is the equation for a line

$$y = mx + b, \tag{8.1}$$

where y is our response (output) variable, x is our control (input) variable, m is the slope of the line, and b is the intercept (the location where the line crosses or intercepts the y axis (where $x = 0$). Once we have this equation, we can use it as our best guess at predicting y from any x value. The mathematical names maybe most familiar to us for these variables are independent and dependent variables. The independent variable is used to explain the dependent variable. The dependent variable is the variable being explained. The independent variable is the variable being controlled (either by being held constant, ignored or intentionally varied); we are measuring or observing the response, which gives us values for the dependent variable. There are several names for these variables, as seen in Table 8.1.

I like using key process (input) variables and response (output) variables when I'm speaking about the experimental variables. The very names remind me of their purpose and their role in the experiment. We are familiar with some of the common models or relationships between variables from our science and engineering classes. For example, Ohm's law describes the relationship between current, I, and voltage, V.

$$V = IR \tag{8.2}$$

TABLE 8.1

Commonly Used Names for Experimental Variables

Variable	Common Names	Explanation
Independent, x	Explanatory variable	Inputs
	Control variable	
	Control parameter	
	Key process variable	
Dependent, y	Dependent variable	Outputs
	Response variable	
	Output variable	
	Response parameter	

The resistance in a circuit (or any portion of the circuit) can be determined by measuring the voltage at various currents and then determining the slope of the line created for the different values of current and voltage. In this case, the current would be our independent variable or control variable and the voltage would be the dependent variable or response variable. Another example is the relationship between stress and strain for an elastic material, where the slope gives us Young's modulus.

$$\sigma = E\varepsilon \qquad (8.3)$$

These linear relationships seem simple, yet we recall from our lab classes that the measured data points, when graphed, do not necessarily form a straight line but contained varying amounts of scatter. In order to establish the relationship between the experimental variables, we drew a line through our (x, y) data points. Our measured data (x, y) points have embedded in them the response to the factor that was changed, random variation, systematic variation, as well as influences from uncontrolled factors. We now come back to something that sounds familiar.

Let's revisit the Input–Process–Output diagram discussed earlier. This is a wonderful tool to use with our experiment to safeguard that the "whole environment" of the experiment is controlled. In reality, a process has four categories of variables, all of which should be examined thoughtfully prior to any experimentation. The variable categories are given in Table 8.2 and illustrated in Figure 8.1. Therefore, if our experiment considers only the control variables and the response variables, we will not have created a reproducible experiment.

The experimental outputs are our response variables. The response variables are those variables that are measured to evaluate the process

TABLE 8.2

Categories of Variables Contributing to Experimental Results

Variable	Designation	Type	Description
Response	Y	Output	Those variables that are measured to evaluate process (dependent variable)
Controlled	C	Input	What we will hold constant during any experimentation (independent variable)
Uncontrolled	N	Input	Anything that cannot be held constant during the experiment (independent variable)
Key Process Variables	X	Input	What we will vary during the experiment (independent variable)

Source: Wortman, B., Richardson, W., Gee, G., Williams, M., Pearson, T., Bensley, F., Patel, J., DeSimone, J., Carlson, D., *The Certified Six Sigma Black Belt Primer*, The Quality Council of Indiana, West Terre Haute, IN, 2007.

performance. The controlled variables are all the inputs that we hold constant during the experiment. Recall that we can create a standard operating procedure as an insurance policy for consistency of these variables. The variables that cannot be controlled during the experiment all fall into the uncontrolled input bucket (noise). We take safeguards to ensure the

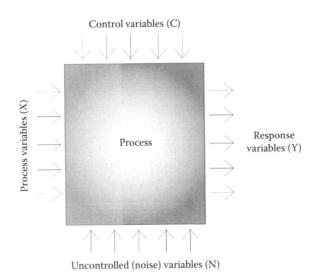

FIGURE 8.1

Illustration of the relationship between different types of variables from an expanded all-encompassing view. (From Wortman, B., Richardson, W., Gee, G., Williams, M., Pearson, T., Bensley, F., Patel, J., DeSimone, J., Carlson, D., *The Certified Six Sigma Black Belt Primer*, The Quality Council of Indiana, West Terre Haute, IN, 2007.)

process is robust enough to be insensitive to these variables; however, this isn't always the case. We may occasionally need to repeat an experiment where an initially uncontrolled variable must be controlled in follow-up experimentation. Finally, the key process variables are those variables that we intentionally vary during an experiment. With these further refinements to our experimental understanding, we can expand the Input–Process–Output diagram as seen in Figure 8.1.

8.4 THE FUN BEGINS...

Once we have data, that's when the fun begins. The data analysis comes with more than a few caveats. However, before walking through a list of safety precautions, let's talk about the tool (regression analysis) that we'll use to analyze our data. Regression analysis, an amazing and powerful tool, will be used to uncover and quantify the complex relationships between control and response variables. As Professor Charles Wheelan warns, "Regression analysis is the hydrogen bomb of the statistics arsenal. ... It is relatively easy to use, but hard to use well—and potentially dangerous when used improperly." He goes on to say, "... regression analysis is arguably the most important tool that researchers have for finding meaningful patterns in large data sets" (Wheelan 2013). As we'll see, there may be some unmeaningful relationships found with regression analysis as well.

Most of us are familiar with one-factor-at-a-time experimentation, and once our data are in a spreadsheet, regression analysis is a simple click away. However, the analysis of the data using regression analysis can be tricky. Like with any tool, it is important to understand its applicability and its limitations so that we are not led astray in our experimental conclusions. Since most of us use computers these days to perform any type of data analysis, we can spend more time with our data, ensuring that we use the computational tools properly.

Although regression analysis can be used to analyze complex relationship between multiple variables, the discussion in this chapter takes only two variables into consideration. In fact, one-factor-at-a-time experimentation cannot (in most cases) account for complex relationships between variables. For example, a certain material may corrode at a certain rate in room temperature water but as the temperature increases, the chemical reaction rate may change. One-factor-at-a-time experimentation would require many experiments to completely characterize and capture the relationship between time, temperature, and corrosion rate for this material. The next chapter will provide tools for experimentation using more complex (multifactor) situations.

We want to use regression analysis to describe (in mathematical descriptors) the "best" linear relationship between the two variables we are investigating. In the scatter plot, we can draw lots of lines through the data. As the name implies, we see a plot with data scattered between the axes. We can eyeball the data and roughly describe the relationship between variable y and variable x. Additionally, we may see that it's possible to draw multiple lines or curves through the data points. One of the advantages of regression analysis is that it allows us to mathematically fit a line that will best describe the relationship between our response (dependent) variable and key process (independent) variable based on the data we have. Always remember that regression analysis will establish a correlation between any two variables that are plotted, even nonsensible data. Tyler Vigen's *Spurious Correlations* is filled with many great examples, such as the correlation between beef consumption and death by lightning strikes or the price of gasoline and the number of lawyers in Texas (Vigen 2015).

Let's look at an example. Let's say we've measured the thermal diffusivity of alumina (Al_2O_3) ceramic at different temperatures (Munro 1997, NIST 2016). The measurements and plotted data are shown in Figure 8.2 along with a curve to show the overall trend in the data. Notice as the ceramic heats up, there is a dramatic drop in the diffusivity. It is possible to begin to build a model with the data points provided. However, because the data points are not linear, we'd need to have a fairly complex model to describe the behavior.

Let's say for our processes, we are very interested in temperatures around 1200°C. We want to build a linear regression model for temperatures between 1000°C and 1500°C. Let's regraph the data but we'll exclude temperatures lower than 1000°C. Figure 8.3 shows the data between 1000°C and 1500°C with a smoothed trend line. From these data, it appears we can build a simple linear regression model. Now, we will need to stress to our audience that whatever model we build from these data is applicable only within this temperature range. For temperatures below 1000°C, the diffusivity dramatically increases. We can extrapolate with our model, but unless we have experimental data, our confidence for temperatures greater than 1500°C diminishes the greater the temperature.

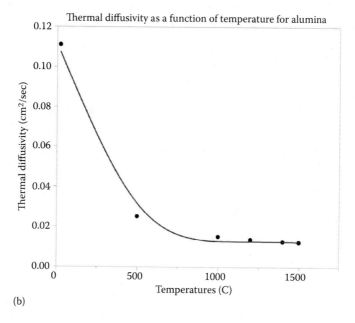

	Temperatures (C)	Thermal diffusivity (cm²/sec)
1	20	0.111
2	500	0.0251
3	1000	0.015
4	1200	0.0136
5	1400	0.0127
6	1500	0.0124

(a)

(b)

FIGURE 8.2

Experimental values (a) and graph (b) of the thermal conductivity of alumina exposed to different temperatures. The line is a smoothed trend line showing the shape of the data points.

A mathematical technique called ordinary least squares is typically used to define our best line. In order for us to really understand how ordinary least squares works, I need to introduce an additional term: *residual*. The *residual* is the delta or difference between the actual response variable data point and the model created from the data (line that is drawn). You are probably thinking that the smaller the residual, the better the fit to the data and vice versa. Now you may be having one of those "AHA" moments, thinking you've got it. You are close. Actually, ordinary least squares fits a line that minimizes the *residual squared*, which accounts for all the outliers

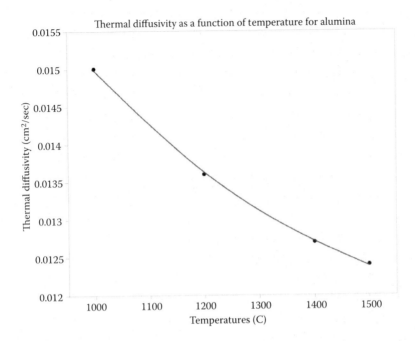

FIGURE 8.3
Experimental values and graph of the thermal diffusivity of alumina exposed to temperatures between 1000°C and 1500°C. The line is a smoothed trend line showing the shape
of the data points.

in the data. Since most spreadsheet software with graphing functionality
uses this method, all we need to do is input our data into spreadsheets,
plot, and fit, then magically, using regression analysis and ordinary least
squares, we have the equation of the line that best describes the relationship between our two variables of interest. Notice that our line (Figure 8.4)
doesn't necessarily touch all the data points; the line does provide the best
model (linear fit) to the data. The fitted line shown in Figure 8.4 gives us the
relationship between the thermal diffusivity and temperature for alumina.
A good graphing program will also provide information about the summary of the fitted relationship.

$$\text{Thermal Diffusivity}\left(\frac{cm^2}{sec}\right) = 0.020 - (5.2034e-6)^*\text{Temperatures}(°C)$$

$$(8.4)$$

Once we have our model, our equation for a line that best fits our data,
$y = mx + b$, we can use it as our best guess at predicting y from an x value.

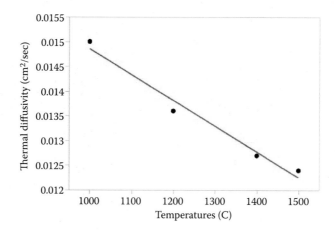

FIGURE 8.4

Measured values of the thermal diffusivity of alumina exposed to temperatures between 1000°C and 1500°C with a linear regression fit to the data.

Can we really use our model to predict *y* values within the range of *x* values that we didn't test? It goes without saying (but I'll say it anyway) that this line can be expected to predict only values within the range of values for *x* that we investigated (see Figure 8.5). The predicted values from interpolation (within the *x* values we tested) will provide values for the dependent

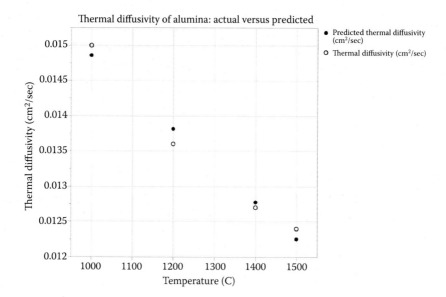

FIGURE 8.5

Overlay of measured and predicted values for the diffusivity as a function of temperatures. The predicted values use the model from the regression analysis.

variable, *y*, that are as good as our experimental data. Can we use it for *extrapolation*? The answer is a strong, resounding maybe. Proceed down the extrapolation path with trepidation and caution. Always keep in mind that, when using a model built for a certain experimental range of independent variables, our model is good ONLY for that experimental range of independent variables. We want to use our models to predict what will happen in other areas of our experimental space. If the model works outside of the original ranges of independent variables, how wonderful! Figure 8.6 shows what happens in our example. The further away from the temperature range of our original model, the worse the prediction of the model. We see at 500°C that the model prediction is somewhat close for the thermal diffusivity. However, the closer we get to 0°C, the more the model diverges from the actual measurements. We might be able to avoid further experimenting in that range. However, if our model doesn't work, we can use this model and our understanding of the prior experimental model to create a new experimental range. In other words, the extrapolated prediction may be used as the starting point for a follow-up experiment.

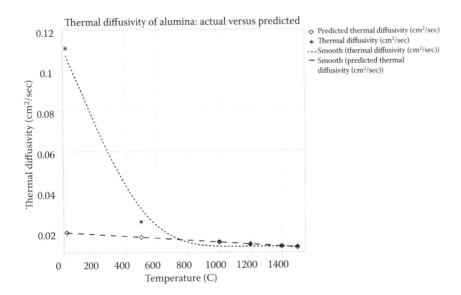

FIGURE 8.6

Overlay of measured and predicted values for the diffusivity as a function of temperatures including temperatures below 1000°C. The predicted values use the model from the regression analysis to extrapolate outside the region of the model.

Now, back to our experimental model. Let's look at our model. Notice whether there is a positive or negative sign in front of the key process variable. The sign tells us whether we have a direct relationship or an inverse relationship. The inverse relationship or negative slope tells us that as the control parameter increases in our model, the response variable decreases. The magnitude of the slope is also of interest in our model. For example, let's say our experimental results show an increase in corrosion rates by a factor of 5.3 for metals exposed to dog urine. Is 5.3 a large or small value? One question we may need to answer is how does this compare to corrosion rates for this same metal not exposed to dog urine under otherwise similar conditions (the population as a whole). The reason the size of the value 5.3 is important has to do with its significance as opposed to the absolute numerical value. The significance is really at the heart of our reason for investigating in the first place. We need to determine if this result is representative. Assuming that we've followed all the rules of experimentation, there is a large enough data set (at least 30 samples), we can use the central limit theorem, the normal distribution and standard error to determine significance (Wheelan 2013). (Refer to a statistics book for other options when working with a smaller data set.)

There is another important part of our model that we need to discuss. Notice the "Summary of Fit" table included in Figure 8.4. Along with our model (the "best fit" equation to our data), a value called R^2 was calculated. The value of R^2 is used to estimate how "good" our "best fit" is to the data. R^2 provides a measure of the variation explained by the regression equation—the proportion of the variance in y attributable to the variance in x. (At this point, we should be wondering about R. R is used to represent something called the *Pearson product moment correlation coefficient*. R is a dimensionless number that ranges from −1.0 to 1.0, inclusively, and reflects the extent of a linear relationship between two data sets. For more information on this topic, I'd recommend having coffee with a statistician or consulting a statistics textbook.) R^2 can vary between 0 and 1.0, inclusively. When $R^2 = 0$, the model that we've built performs no better than the mean at predicting the relationship between our experimental variables. When $R^2 = 1.0$, the model predicts this relationship between the two variables exactly. Typically, we will find that the R^2 value is somewhere in between these two extreme values.

Remember the discussion in Chapter 1 on the caution about collapsing correlation and causation. Refer back to Figure 8.4. In this case, we see the $R^2 = 0.977$. It is probably reasonable to say that the temperature change is

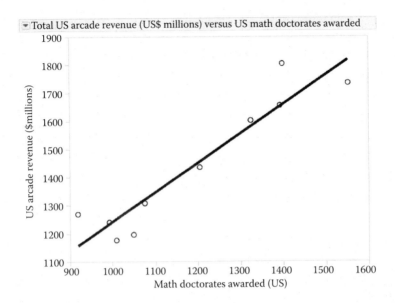

FIGURE 8.7

There is an R^2 correlation of 0.89 between US arcade revenue and math doctorates awarded between 2000 and 2009. (From Vigen, T., *Spurious Correlations: Correlation Does Not Equal Causation*, Hachette Books, New York, 2015.)

in part a source of causation for the diffusivity change in Al_2O_3 due to the strong correlation we see between the two variables *and* our knowledge of subject matter. Knowledge of subject matter tells us (recall from freshman physics) that temperatures of radiating bodies actually scale as T^4. However, we saw that temperature corrections over small ranges can be treated linearly. An $R^2 = 1.0$ simply means correlation; it doesn't imply causation. Without knowledge of subject matter, we are wandering around in the dark. Take the example in Figure 8.7, which shows the correlation of math doctorates awarded and arcade revenue, though we probably cannot build a case for causation.

We've just walked through a discussion about linear regression analysis using ordinary least squares and built a simple but beautiful linear model for our experimental data. However, there are many times when the relationship between the two experimental variables isn't going to be linear or best described by a line. This doesn't mean that linear regression isn't applicable. We've covered a simplified case of the general form of regression. Once we are comfortable and confident with the simplified case, there are references available for more sophisticated model development later in the book.

8.5 KEY TAKEAWAYS

The most common experimental strategy in use in the physical sciences today remains one-factor-at-a-time experimentation. It is a good way to get our feet wet, so to speak, in experimental problem solving. If there are adequate resources available for experimentation, this tool is the by far the most intuitive to gain a basic understanding of experimentation and model building from our data.

The one-factor-at-a-time technique is a perfectly good experimental strategy, but it is limited. A one-factor-at-a-time experiment may be what is needed when we know that the variable interactions are not complex or if we have a large resource pool so the number of experiments that we can perform is not limited. Limited resource situations with known or suspected complex interactions are the occasions when designed experiments of higher order are needed in our experimental toolbox. For these situations, we'll need Experimenting 201 in Chapter 9.

P.S. We'll move to more complex experimentation in Chapter 9. When complex relationships exist between variables, we need to use more sophisticated techniques than one-factor-at-a-time experimentation. I learned about complex relationships very early on the farm. My father was interested in the tomato yield each year. Every day, I watched and watered the plants at roughly the same time of day and recorded whether we had tomatoes or not. I dutifully counted the tomatoes we harvested from each plant and the harvest date in the family garden for two consecutive summers. At the end of the each summer, I proudly showed my father this fancy prediction capability that I had learned in school and proclaimed that I could tell him almost exactly when the tomatoes would start producing and the yield rate over time. This little experiment taught me a valuable lesson in experimentation. Reflect back on your own experimental experiences, what lessons have you learned with one-factor-at-a-time experimentation?

REFERENCES

Boorstin, D. 1983. *The Discoverers*. New York: Random House.
Dolnick, E. 2011. *The Clockwork Universe: Isaac Newton, the Royal Society & the Birth of the Modern World*. New York: HarperCollins.
Munro, R. G. 1997. Evaluated Material Properties for a Sintered alpha-Al_2O_3. *Journal of the American Ceramic Society* 80:1919–1928.

NIST. 2016. National Institute of Standards and Technology Ceramic Data Portal contains experimental data. http://srdata.nist.gov/CeramicDataPortal/Pds/Scdaos.

Vigen, T. 2015. *Spurious Correlations: Correlation Does Not Equal Causation.* New York: Hachette Books.

Wheelan, C. 2013. *Naked Statistics: Stripping the Dread from the Data.* New York: W. W. Norton.

Wortman, B., W. Richardson, G. Gee, M. Williams, T. Pearson, F. Bensley, J. Patel, J. DeSimone, and D. Carlson. 2007. *The Certified Six Sigma Black Belt Primer.* West Terre Haute, IN: The Quality Council of Indiana.

9

Experimenting 201

Be a detective, not a lawyer. ... A lawyer's job is to prove or persuade. A detective's job is to find things out.

John Sall

There are times when a one-factor-at-a-time experimental approach will not allow us to accurately explore the process space of interest. In these cases, we need more sophisticated tools. Recall my tomato harvesting example from the last chapter, in which I built a model to predict when we could expect our tomato plants to yield ripe, delicious, ready-to-eat tomatoes and the yield rate during the balance of the plants' lifetimes. I'm sure any gardeners reading this are shaking their heads at my youthful naivety. The next summer, the harvest date was much later and the yield rate was wildly different. I hadn't accounted for rainfall, temperature, etc. Like with this example, in real life, the interactions between variables are complex, which means our experiments must become increasingly sophisticated. This is where designed experimentation techniques come in handy. With designed experimentation, we are again looking at *intentional variation* where we've controlled for *unintentional variation*, *systematic variation*, and *random variation*. However, unlike one-factor-at-a-time, we will now vary multiple factors at one time in a strategic, designed approach.

9.1 COMPLEX PROBLEMS

In today's world, we want to measure everything, from the number of steps we walk to protein levels in our blood. Luke Dormehl, in his book

The Formula: How Algorithms Solve All Our Problems ... and Create More, relays the story of one such person who regularly, quantifiably monitored his health only to observe a certain protein, indicative of infection, increasing. Bringing his data to his personal physician, he was rebuked for coming in with data and not a health problem. Weeks later, the man was having surgery to remove his appendix (Dormehl 2014). We've come a long way from leeches sucking out the "bad" blood, but maybe not as far as we like to think.

It wasn't until the 1950s that medicine began its transition from art to science. In the physical sciences, we have Galileo to thank for ensuring that investigations in physics "will never be the same" (Sobel 2000). Galileo stopped looking for why natural phenomenon happened and began observing and measuring (repeatedly) what was actually happening in nature. In medicine, we have prisoner of war Dr. Archie Cochrane to thank for introducing a scientific approach. The first evidence we have of statistical experimentation in medicine was from the work of Dr. Cochrane during his World War II imprisonment. He performed random control trials on fellow prisoners (Sur and Dahm 2011). The expansion and benefits of randomized control trials were further developed by Drs. Thomas Chalmers, Ian Chalmers, and Murray Enkin in the decades of the 1950s to 1960s. The physicians showed that even medicine is susceptible to both evidence selection and bias. For most of us today, we can't imagine what a radical shift this actually was, and it didn't happen overnight. Evidence-based medicine was actually coined in 1991. Seriously, you read that correctly. In 1991, Dr. Gordon Guyatt introduced a new method for bedside teaching of residents called "Scientific Medicine" later changed to "Evidence-Based Medicine" in an editorial he authored for the ACP Journal Club (Guyatt 1991). Quoting Dr. Deborah Kilpatrick, chief executive officer of Evidation Health, "Controlled clinical trials and formalized, evidence-based recommendations as to how medicine should be practiced is a fairly recent phenomenon" (GE 2016). We are actually living in the midst of this revolution—or maybe it would be better to call it a paradigm shift—in the way scientific data will be used in the practice of health care. The algorithms, formulas, and/or models that come out of the systematized, scientific approach to the analysis of the volumes of data collected about us are already impacting our lives. This is evident in Google's, Amazon's, and Facebook's use of what we click on and even how long we hover over a particular screen, or from Apple's iPhone, or our Fitbit tracking where we go and how many steps it takes to get there. The building of models from collected data is a key piece of statistical experimentation in any multivariate problem—whether in the physical sciences, medicine, shopping, or searching. Designed experimentation is an important tool to have in our toolbox anytime we suspect that there is more than one variable that can influence the results.

The problems faced in medicine, Internet searches, shopping, and science are complex and multidimensional. The many variable problems of our real world require that we be able to experiment and analyze the results for more than one factor at a time. One-factor-at-a-time can be a great way to begin experimenting. However, this type of experimentation can be limiting. If we visualize the possible experimental variables as forming a multidimensional space and experimentation as a means of exploring that space, it's easy to see that we could spend years experimenting with all the possible combinations of factors.

The human body is a perfect example of an incredibly complex multivariate problem. In order to study the effect of all the key process (input) variables (KPVs) on the response (output) variables we are interested in, there are prohibitively too many confounding variables for accurate one-factor-at-a-time experiments. Additionally, if our goal is to identify a stable optimum condition, it is highly unlikely that we will achieve this with one-factor-at-a-time experimentation. In a manufacturing process, we may never find a stable operating condition with one-factor-at-a-time experimentation. Conditions are always changing. For example, equipment has mechanical components that fail with time and tools wear out with use. In our lives, there may be interactions between variables that alter the behavior of the response variables. Think of sleep, exercise, stress, and diet. These interactions are almost impossible to quantify with one-factor-at-a-time experiments. This might lead to wrong, misleading, inconclusive, or suboptimal experimental results. Without an in-depth understanding of variable interactions, mysterious or inexplicable effects may impact the results. In the end, time and effort may be wasted through experimenting with the wrong variables or running too few or too many experiments. Designed experiments overcome these problems through careful planning.

Design of experiments refers to the methodology of varying a number of process (input) variables simultaneously, in a carefully planned manner, such that their individual and combined effects on the response (output) variables can be identified. In the literature, the acronym DOE refers to either design of experiments or designed orthogonal experimentation. This designed orthogonal experimentation refers to a particular type of designed experiment that is orthogonal. What is orthogonal? A design is orthogonal if the main effects and interactions in a given design can be estimated without confounding the other main effects and interactions. A full factorial is said to be orthogonal because there are an equal number of data points under each level of each factor.

Both in lab and in industrial environments, experimenters will have limitations on their most valuable resources: money and time. Although we don't often like to talk about these mundane issues in texts, they are the reality nonetheless. With designed experimentation, many variables or factors can be evaluated simultaneously, making the designed experimentation process economical and less interruptive to normal lab or industry operations. With proper planning and execution, this methodology can save money and time as compared to more traditional one-factor-at-a-time experimentation. Designed experimental strategies allow us to draw conclusions from fewer experiments. The rigorous statistical backbone of designed experimentation highlights important variables or factors and distinguishes them from less important ones. With the important factors identified, additional resources can be directed toward an increased understanding of these factors. If by some chance critical variables or factors have been overlooked in the experiment, this will show up in the results. The snubbed variables can be investigated in follow-up experiments. The previously tested variables may be set at more optimal conditions and then tested under the new experimental conditions, thereby providing verification and inspiring confidence in the results. However, as we see later in the chapter, when interactions are significant, proceed carefully when making assumptions.

As with traditional one-factor-at-a-time experimentation, noise or noise factors may be an issue no matter what precautions we take. Sometimes, noise factors cannot be directly influenced, instead other input factors can be controlled to make the output less sensitive to noise. Whether running an experiment with 1 variable or 10, the more information that is known about noise and nuisance factors, the more confident we can be in our results.

An in-depth statistical knowledge is not necessarily required to learn or benefit from designed experimentation. The statistical analysis can be performed using readily available software. There are many excellent statistical packages available with varying price tags. I have worked with JMP for more than 20 years and personally prefer the combined data exploration, analysis, and experimental design capabilities. The companies that sell these software programs provide excellent technical support both in publications as well as in person either over the phone or in training. However, knowledge of statistics and statistical experimentation methods will minimize opportunities for rookie mistakes. The commercially available software packages should never be treated like "black boxes."

9.2 ESTABLISHING THE EXPERIMENTAL PROCESS SPACE

The experimental process space will be defined by the process variables we select. When selecting our process variables, we should start with our best *Input–Process–Output* diagram. Create an Input–Process–Output diagram for the process and make sure it is thorough. We will want to have one or more people knowledgeable about the subject review our Input–Process–Output diagram prior to narrowing down variables. A technical review with a second set of eyes will help ensure the thoroughness in the diagram. Identify the key independent and dependent variables and determine if these factors are controlled (C), noise (N), or intentionally varied (X). Process variables include both inputs and outputs (factors and responses). Selecting these variables may be best done as a team effort and should include all responses. We can always remove irrelevant responses that do not matter to our objective.

Here's where our visualization skills meet science. We want to imagine the process space as a physical space. Whether we are running an experiment with one, two, three, or more levels for a variable, our experiment will create a topographic map of the process space we define. The number of levels determines the amount of detail within the process space. The extreme (high and low) levels that we choose define the process space. For example, I might run an experiment where I vary the salinity of water and measure the change in mass on copper after 10 days. This one-factor-at-a-time experiment is simple to visualize graphically with a simple two-dimensional graph with some measure of salinity along the x axis and mass change on the y axis. What happens when I add the measurements after 15 days and 20 days? I could make two graphs and put them beside one another for comparison (Figure 9.1a). Another option is to graph both sets of data on the same two-dimensional graph (Figure 9.1b). A third option would be to think of this as my process space with % salinity along one axis and time along the other axis (Figure 9.2). Note that it may be a little more challenging to compare data that aren't plotted on the same graph. The graphs for the second and third options are much more effective in visualizing the process response map.

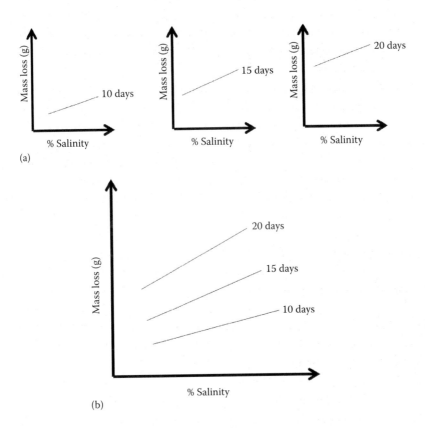

(a)

(b)

FIGURE 9.1
Two ways to visualize the relationship between variables: (a) Three graphs displaying data versus (b) all data on a single graph.

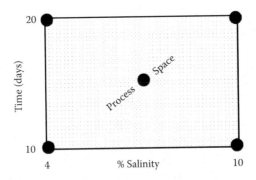

FIGURE 9.2
A third way to visualize the process space that captures the relationship of different variables that easily extrapolates up to four dimensions.

9.3 SELECTING A DESIGN

Once we have an idea of our process space, before actually beginning experimentation, we will need to select a design for our experiment. When we are designing our experiment, either in a software program or by hand, we want to take a chance but not be impractical in choosing the process space. We want to assign levels to each independent variable in light of our knowledge of the process, equipment, resources, etc. It is critical that we *not* throw out our experiences in exchange for a computer program. "The computer told me to do it" is not a valid excuse. It is beneficial, even for experienced experimenters, to have another person critically review the outline of the designed experimentation.

An experiment is typically designed with the goal in mind. Keep that goal in mind during the design selection. There are several primary reasons we want to run designed experiments with multiple variables. Two of the most common experimental objectives for scientists and engineers are comparing and screening. Table 9.1 lists several common objectives for designed experiments. Notice in Table 9.1, even one-factor-at-a-time experimentation can be improved through design. When the goal is comparing, we have several factors under investigation and our primary goal is to determine which factor or factors are "significant" or the "most significant." In this case, we need a comparative design solution. The goal of screening experiments is to screen out the important experimental variables. Screening designs allow us to evaluate a large number of experimental variables with very few experimental runs. Therefore, typical screening experiments involve two-level designs with varying degrees of fractionalization. A *full factorial screening design* will have us run all combinations of the process (input) variables (X). A *fractional factorial screening design* experiment is a fraction of a full factorial experiment. A fractional factorial screening design allows us to quantify the changes occurring in

TABLE 9.1

Common Design Objectives and Guidelines for Selecting a Design

Objective/ No. of Factors	Comparing	Screening
1	Randomized one-factor-at-a-time design	
2 or more	Randomized block design	Full or fractional factorial

the response (output) variable (Y) of a process while changing more than one process (input) variable (X). With a fractional factorial, there is no need to run every combination of experimental conditions. The fractional factorial screening design uses confounding to consume fewer resources. *Confounding* means that the value of a main effect estimate comes from both the main effect itself and a contamination of higher order interaction terms.

Of course, with a fractional factorial screening design, there are trade-offs. The advantage is that we run fewer experiments in less time and with fewer resource requirements than with a full factorial screening experiment. The results from a fractional factorial screening design will be average responses associated with multiple factors. We must be very careful when interpreting the results of these fractionalized screening experiments. Just because a factor is not highly significant in a fraction-alized design does not necessarily mean that it is not significant. The experimental design itself is critical in ensuring that significant effects aren't missed. However, the fractional factorial screening design has less power to quantify the interactions between the process (input) variables because of the confounded effects. Many today think fractional factorials are outdated. Although the terminology is still used, it is possible to work through custom design and definitive screening designs to create efficient designs without having to work through the traditional steps of fractional factorials. I've included fractional factorials for comparison only.

Besides comparing and screening, there are other experimental objectives that involve more advanced designs and concepts. Designed experiments intended for mapping (again the cartography comparison) allow an experimenter to discover the shape of the response surface (topography) under investigation. These experiments are aptly named *response surface designs*. The response surface design will fully explore a process window. Typically, this type of design is used to improve or optimize a process space or troubleshoot in a well-understood process space. This type of experimentation is best performed with a well-understood Input–Process–Output diagram and is rarely, if ever, the initial experiment performed. An experiment intended to fully map out a process space would be performed as a follow-up experiment to other experimentation. Response surface designs are most effective when there are fewer than five process (input) variables (X). These designs are resource intensive, requiring at least three levels of every process (input) variable (X). However, quadratic models are generated for each of the response (output) variables (Y).

There are two other experimental objectives that we may encounter. Optimizing responses when process (input) variables (X) are proportions of a mixture may necessitate a *mixture design*. A mixture design will aid in optimizing the process (input) variable proportions (X's) to maximize or minimize the response (output) variables (Y's). The other type of design that might be of interest is a *regression design*. In order to model a response as a mathematical function (either known or empirically determined) of a few continuous process (input) variables (X's), to obtain model parameter estimates, a regression design might be a good option. These designs are not covered here, but Chapter 12 provides references to more advanced texts on designed experimentation.

Once we have objective of our experiment nailed down, selecting the type of design is critical. The easiest experimental designs to understand are two-level designs. Two-level designs are simple, typically economical, and give most of the information required to go to a multilevel response surface experiment if one is needed. Two-level design is really a misnomer. In most cases, we include some additional points during the experiment (at the center is typical) to check for nonlinearity. By adding a center point, we are essentially adding a third level.

I want to add a note here about some of the more complex designs. These designs are named after great statisticians (Taguchi, Plackett, Burman, etc.), and it sounds really sexy (yes, that is a matter of perspective) to proclaim to the world "I'm running a Plackett-Burman blah blah blah." I will be honest; I've never run a Taguchi or Plackett-Burman design. The confounding pattern of the variables is so complex that even the simplest designs are difficult to understand. My advice is: if you don't understand it, just avoid it or find someone who does and have them keep explaining it until you get it. Dr. Khorasani and I presented a paper at a Sematech conference where we proposed a method to model the experiment before beginning to ensure that the design would return the expected results. This method is a safeguard against complex, confounding patterns that aren't easily interpreted (Buie and Khorasani 1998). Any model with an extremely complex and highly confounded interaction is rarely useful. Remember our goal in problem solving is not to develop complex models but to translate the models that we develop into a solution for the problem we are attempting to solve.

The choice of the type of design depends on the amount of resources available (time, money, samples, etc.). It is a good idea to choose a design that requires somewhat fewer runs than the budget permits, so that additional runs can be added to check for curvature and to correct any experimental mishaps. There are a few decisions that we will want to make prior to the design selection that should help make our design selection easier. Another consideration in choosing the design is the difficulty with which an experimental run can be changed. For example, biologist and mathematician, Sir R. A. Fisher, developed *split-plot designs* for use in agricultural experiments (Fisher, 1925). A split-plot design naturally blocks the experiment such that the blocks are experiments. Dr. Bradley Jones and Professor Christopher Nachtsheim have published a paper with details of split-plot design motivation (Jones, 2009). Other examples can found in *Statistics for Experimenters, 2nd Edition* (Box, 2005). The field of designed experimentation is continuing to develop. Most industrial and engineering experiments are split-plot designs, which makes it a valuable technique for scientists and engineers.

The very first things we will want to identify are the response (output) and process (input) variables that are important to us. The response variables (Y's) are the variables that show the observed/measured results of our experiment. The process variables (X's) are the independent variables that have some type of effect on the response (output) variables (Y's). The levels or conditions that we use for input factors will determine the response(s) that we eventually will measure. Both the response (output) variables and the process (input) variables can be qualitative or quantitative. Quantitative measurements (numeric and continuous) tend to be preferred by most engineers and scientists; however, there are times when the only response possible is qualitative. Quantitative measurements are those that give us a numeric value, a quantity. Qualitative inputs or responses (characters and nominal) have different properties or attributes. For example, let's say we are interested in whether we have created a leak-free join. Leak detectors allow us to determine whether a leak is present or not. Our responses might be a qualitative yes (a leak is measured) or no (a leak is not measured). However, rather than leak-free join, we may be interested in creating and holding a vacuum. If our leak detector is sensitive enough, we might be able to measure the actual leak rate and have a quantitative (continuous numeric value) result to use in our analysis. There may be times when we are working with discrete numeric input parameters rather than continuous parameters. Discrete numeric values are more desirable than qualitative data but less desirable than continuous data.

Once we have selected the response and input variable(s) or factors that we are interested in, the next step is to decide on our process space. The process space is determined by the extremes or highest and lowest values that we are interested in exploring. For example, let's say we are interested in the effect of bath temperature and time on the nickel electroplating thickness onto aluminum. In a screening experiment, the process space will be determined by the highest and lowest values chosen for temperature and time. If we decided to vary the time the parts are in the bath from 10 to 20 seconds and vary the temperature from 10°C to 40°C, these settings will determine the outer range of the experimental process window. With the range defined, we will want to select the number of levels within that range we want to run for the experiment. The most common are two or three levels or values. For example, the low value for time would be 10 seconds, the high value would be 20 seconds. For the third value, we might choose 15 seconds. The number of values is, in part, determined by the resources (parts, time, etc.) that we have available to experiment upon. The more values we choose to experiment upon, the more confidence we can have in the process space.

With a screening design, the number of experiments we would like to run will be determined by the resources we have available to dedicate to the experiment. The simplest case is known as a *full factorial*. A full factorial screening experiment will use all combinations of all the input values that we selected. Depending on our resources, a full factorial might be very expensive. An alternative screening design is a *fractional factorial*, which can significantly reduce the number of experiments performed.

A full factorial designed experiment consists of testing all possible combinations of process (input) levels. The total number of different combinations for k factors at two testing levels is $= 2^k$. For example, in our experiment with two factors and two testing values each, there will be a total of $n = 2^2$ or 4 combinations. This allows us to create a matrix of experimental runs.

The advantage of testing the full factorial is that we obtain information on all main effects plus all interaction effects. The *main effect* is an estimate of the effect of a factor independent of any other factors. Let's take the previous electroplating example with two input factors and two values. The main effects would be time and temperature. An *interaction effect* occurs when the effect of one input factor on the output depends on the level of another input factor. The interaction effect would be the effect of the interaction between time and temperature, written as time * temperature. These effects are key to the type of model that we build from our experiments. With $n = 2^2$ or 4 experimental runs in our plating example, the model would include time, temperature, and time * temperature.

Let's examine a more complex screening experiment. We want to understand the effect of a new piece of hardware on the etch properties of a TiW film. We decide that we are interested in five input factors: pressure, power, BCl_3 gas flow, N_2 gas flow, and CF_4 gas flow. These five input factors are examined at two values each. A full factorial screening experiment would have us run $n = 2^5$ or 32 experiments. We would then know the effect of each of the following: pressure, power, BCl_3 gas flow, N_2 gas flow, and CF_4 gas flow, pressure * power, pressure * BCl_3 gas flow, pressure * N_2 gas flow, pressure * CF_4 gas flow, power * BCl_3 gas flow, power * N_2 gas flow, power * CF_4 gas flow, BCl_3 * BCl_3 gas flow, BCl_3 * N_2 gas flow, BCl_3 * gas flow, and CF_4 * N_2 gas flow. This is quite a cumbersome and complex model.

Interaction effects are easily identified in a full factorial experiment but can be lost in a fractional experiment. Recall the loss of interaction effects is called confounding. When interaction effects are confounded with another interaction effect or a main effect, there is no way to distinguish which effect is actually important, or maybe both are. There are rare occasions where an interaction is more important (or more significant) in an experimental space than the main effect is. In the TiW metal etch example with five input factors, we have several options for fractional factorial designs. There are two primary choices of fractional screening designs: $n = 2^{5-2}$, or 8 experiments, and $n = 2^{5-1}$, or 16 experiments. The design with eight experiments will resolve only three of the main effects. All other effects will be confounded. The $n = 2^5$ or 16 experiments will resolve all five of the main effects only. All interactions will be confounded with one another. The more fractionalized the factorial experiment, the more confounded the factors. If we have any resource constraints at all, the selection of design is critical. Designing the experiment and selecting the number of runs strategically are essential. See references in Chapter 12 on how to strategically select the best design.

In all likelihood, choosing the process space and selecting the design to use will be an iterative process. In an effort to avoid factor settings for impractical or impossible combinations, test each run in the experiment ahead of time. Run "dabbling experiments" where necessary to debug equipment or determine measurement capability. This will allow us to further develop experimental skills and additionally get some preliminary results.

Randomize experiments as much as possible. The statistical software program can randomize the runs. This may or may not be an important step in our experiment. Obviously, it doesn't make any sense to talk about *randomization* if our experiments are being performed in parallel, as is the case with the change in mass versus % salinity. Likewise, if our experiment were observing the change in mass over time for different % salinity, we wouldn't want to randomize to the point where our $t = 20$ day experiment came before our $t = 10$ day experiment. There may be other less obvious cases, as in the case where there is some time-intensive equipment change that needs to happen for a certain number of the runs. We may want to lump those together so as not to completely randomize the setup variable. An easy example of this is temperature. If one of my variables is oven temperature with temperatures of 200°C, 300°C, and 400°C, I might want to run all the 200°C experiments before increasing the temperature. Here, I've described a prototypical split-plot experiment with an easy-to-change factor and a difficult-to-change factor. A similar example can be found in *Statistics for Experimenters, 2nd Edition* (Box, 2005).

Another cautionary approach is to analyze results periodically. With experiments that are expensive either in time or resources, periodically confirming that everything is on track and performing as expected may provide us with an early indication that something has gone wrong.

Finally, let's not put all our eggs in one basket! It is often a big mistake to believe that one big experiment will give the answer. A more useful approach is to recognize that while one experiment might give a useful result, it is more common to perform two, three, or more experiments before a complete answer is attained. An iterative approach is usually the most economical. Among engineers, the whole idea of design–run–evaluate–optimize is understood as a part of the iterative approach to designed experimentation.

9.4 RUNNING THE EXPERIMENT

The second part of the process is to execute our designed experiment. While our experiment is running, we should stay as close by as we can. Some experiments take place over weeks, and it isn't possible to be present the whole time in these cases, but it is essential that we own each step in the experimental plan and execution. Watch out for process shifts and drifts during the run and avoid unexpected changes, but allow some extra time and resources for the unplanned. We want to capture all that occurs during our experiment; even the most mundane observations or subtle changes might be important in the end. While the experiment is running, be sure to capture and document all process data to maximize available information during analysis.

9.4.1 Experimental Example

Let's walk through an example and review what we've covered up to this point. We've been asked to develop a cleaning process for a titanium rod to be used in our new robot. The housing for the rod has been found to be sensitive to ionic contamination. For example, trace amounts of chlorine or sodium ions may result in corrosion of the copper parts, which would affect the performance of the product. Our experimental process variables are as follows:

1. Time in the ultrasonic bath, with a range of 5 to 15 minutes
2. Cleaning solution, with a range of 4% to 10%

We can think of the experimental process space as a rectangle (Figure 9.3a). We are attempting to map out the space. Remember, it's like we are creating a topographical map. With every additional factor, we add a dimension to this visual picture. For our experiments, we will be mapping out a flat rectangle. However, if we add a third factor, for example, rinse time, our process space would look like a cube.

Each of the experiments is called a "run" and refers to the exact experimental combination of factor settings. To conduct the experiment in a way that reduces potential bias from factors not included in the test matrix (e.g., type of raw material, operator, time of day, etc.), we should use a randomization procedure. We want to conduct the runs in a random order. There are a lot of different randomization strategies, and most software packages will generate the randomized sequence. There are times when it isn't possible or the cost is prohibitive to randomize experiments. Examples include when an experiment requires extensive hardware setup for a large experiment.

The center point provides us with a measure of nonlinearity or curvature within our topographical process map (Figure 9.3b). *Curvature* refers to nonstraight line behavior between one or more factors and the response. Curvature is usually expressed in mathematical terms involving the square or cube of the factor. For example in this model, $Y = B_0 + B_1 X_1 + B_{11}(X_1^* X_1) + \varepsilon$, the term $B_{11}(X_1^* X_1)$ describes curvature.

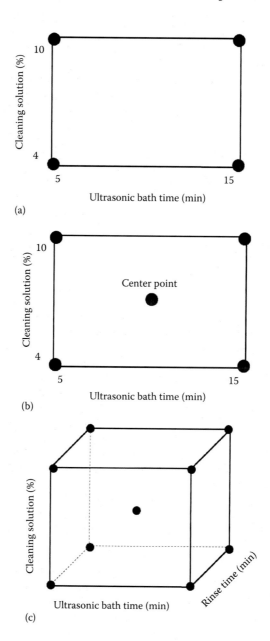

FIGURE 9.3
Illustration of (a) the process space under investigation, (b) the process space under investigation with center point added, and (c) the process space under investigation with a third variable added.

We are also interested in the level of confidence we can put in our measurements. One way to determine repeatability is to replicate one or more of the experiments by performing the same experiment multiple times. Repeated trials, or *replicates*, are conducted to estimate the pure trial-to-trial experimental error or random error independent of any lack of fit error.

9.5 ANALYSIS

The final and most important step is to analyze and interpret the results. We want to confirm that the data are consistent with the experimental assumptions and that the results are consistent with what we know about the subject. For example, if I reduce the amount of heat that I applied to water below a certain temperature, I don't expect the water to boil. The findings may lead to further runs or additional designed experiments. It is important that we take the time and learn all that we can from the results and have others ask us questions about the process as well as the results. Software packages such as JMP that allow us to design the experiments will help with all the calculations and graphs needed for the analysis. However, generating graphs using advanced statistical software is only half the battle; the analysis and conclusions are still up to us. We will still need to translate the graphs and models into the physical realm to answer a question or solve the problem we were interested in.

It is important that we draw conclusions from our analysis of the experiment. We must ask ourselves at each step: "Do these results make sense with what I know about the subject?" A "surprise" result doesn't mean that something has gone wrong, but it is important to verify our findings. Verification can be achieved by replicating runs or whole experiments. Depending on our experimental objective, we may want to proceed with further experiments.

The prior chapters all build on one another. All the information covered in the preceding chapters is important and applies here. A properly executed experiment will ensure that the right kind of data is collected and that there are enough data to meet the objectives of the experiment.

Also, we may want to avoid using responses that combine two or more process measurements. For example, a critical response in thin film etch processes is uniformity. Uniformity can be calculated in a number of ways, but all involve several calculations. Typically, we need to measure the pre-etch film thickness because the incoming film will have some topography and varying thickness. Once the etch is complete, we measure the post-etch film thickness. The difference in the film thickness will give us the etch depth or film removed, which is calculation number one. Because we are interested in a uniform etch, we need to make these measurements at multiple locations. The crudest estimate would be two locations—at the center and edge. Typically, the measurements are performed using a pattern that will cover multiple locations on the surfaces. The calculation of uniformity might consist of anywhere from two points to tens, hundreds, or thousands of measurement points. The most complex calculation for uniformity uses the *coefficient of variation* (a measure of relative variability). The coefficient of variation (CV) is determined by taking the ratio of the standard deviation and the mean. This calculation provides a number that represents the relative variability of the etch depth on the silicon wafer. Embedded in this simple estimate for the uniformity is first a difference calculation (post-thickness minus pre-thickness), then an average to get the mean and standard deviation calculation. The calculated value for the *coefficient of variation* is far away from what we actually measured. This may be unavoidable, but the closer we stay to the actual measurements, the more accurate our experimental results.

We have run a designed experiment in a controlled manner. Now, we want to measure the effect of multiple key process variables (X's) on a particular response variable (Y). In our case, we are looking at two variables. Because we are doing a screening experiment, by definition, our goal is to create a model for the response variable that allows us to identify how important each of our key process variables is to the response variables of interest.

If we want to design an experiment to test the effect of one of our key process variables on our response variables, what would we do? We'd design our experiment so that we tested our key process variables at multiple levels and then measure the response. For example, if we ask the question: Does the temperature of salt water make a difference in the corrosion rate of steel? We might decide to look at samples of water from the Pacific Ocean or we might decide to make our own solution of salt water so that we could more accurately control the properties of the water and the % salinity. We might do a bit of research and see that ocean, slough, and bay temperatures vary. Once we ran our experiment, we would be able to plot the data on a line graph with temperature as our *x* variable and mass change as our y variable. From this, we could create an equation that would model the data for the temperature range in our experiment. Similarly with multivariable experiments; however, the math just becomes more cumbersome the more dimensions we add. With a 2^2 or 2^3 full or fractional factorial, the analysis could easily be done by hand for the experiments. The more experience we gain, the more complex the experiments where performing the analysis by hand isn't practical and really adds no value to the solution. However, it is important for us to understand the analysis so that we understand the limitations.

9.6 CODED VALUES

I want to walk through an example of solving a simple designed experiment by hand, but first, let me introduce a common method to simplify the math by introducing coded or scaled values. Scaled values are simplified values used in the analysis to simplify building the model and create standardized, scaled units for all key process variables (inputs). Coding will allow us to work with 1 and −1 rather than the actual values of 5 and 15, for example. Even though the equations look ominous, this process is simple. To find the coded values:

$$KPV1_{actual} = \left(\frac{KPV1_{Hi} + KPV1_{Lo}}{2} \right) + \left(\frac{KPV1_{Hi} - KPV1_{Lo}}{2} \right) * (KPV1_{coded})$$

where

$KPV1_{coded} \equiv$ code value for key process variable1,

$KPV1_{actual} \equiv$ actual value for key process variable1,

$KPV1_{Hi} \equiv$ actual high value for key process variable1, and

$KPV1_{Lo} \equiv$ actual low value for key process variable1.

For example, consider key process variables time from our titanium rod cleaning example earlier, in which the high value was 15 minutes and the low value was 5 minutes. If we substitute these values into the previous equation, we get:

$$KPV1_{Hi\,(actual)} = 15°C = \left(\frac{15°C+5°C}{2}\right) + \left(\frac{15°C-5°C}{2}\right) * (KPV1_{Hi\,(coded)})$$

$$15°C = 10°C + 5°C * (KPV1_{coded})$$

$$KPV1_{coded} = 1.0$$

Similarly for the low value we find:

$$KPV1_{Lo\,(actual)} = 5°C = \left(\frac{15°C+5°C}{2}\right) + \left(\frac{15°C-5°C}{2}\right) * (KPV1_{Lo\,(coded)})$$

$$5°C = 10°C + 5°C * (KPV1_{coded})$$

$$KPV1_{coded} = -1.0$$

We can repeat this coding procedure for the cleaning solution and the rinse time. A full factorial designed screening experiment for these three input factors will scale the high and low values for the clean solution, 10% and 4%, to 1 and −1 and the rinse time from 5 and 1 minute to 1 and −1, respectively.

9.7 FULL FACTORIAL EXAMPLE

Unless you are taking an exam in statistical experimental analysis (or your Six Sigma Black Belt Exam), you will most likely never need to perform the analysis of a designed experiment by hand. I will do it here to allow you to see what the software is doing. Afterward, we'll compare the result to that of JMP.

Let's pick back up on our cleaning experiment, but we are going to add another variable to the experiment, rinse time. Cleaning has become more and more of an issue in high-tech industries where parts are used for atomic and molecular level etch and/or deposition or in a vacuum chamber. The three key process variables (inputs) that we want to vary in the experiment are ultrasonic bath time, cleaning solution, and rinse time. As the engineers responsible for the experiment, we wish to identify the key process variables affecting the removal of the trace amounts of sodium ions from the parts. We decide to run a full factorial experiment because it is suspected that there may be important interactions between the process input variables that may impact the quantity of sodium ions on the parts. We want to determine the effect of all three factors and their interactions; therefore, a 2^3 full factorial must be run. We establish high and low values based on our existing knowledge of the process and equipment. The values and input factors are shown in Figure 9.4.

A 2^3 screening full factorial will contain eight different experimental runs, which JMP will generate. Our experiments can be seen in Figure 9.5. Normally, we want to perform the experimental runs in a randomized order. JMP will allow randomization of the runs; however, for illustrative purposes, the runs are sorted in a pattern from left to right.

For simplicity, we will not deal with any center points or replicates in this example. Also, to make the analysis by hand simpler, I'd like to use the scaled or coded values. In this case, our experimental runs for analysis would look like Figure 9.6. I am doing this only to illustrate what's behind

Factors

Continuous	Discrete numeric ▾	Categorical ▾	Remove	Add n factors	1	

Name	Role	Values	
◢ Ultrasonic bath time (min)	Continuous	5	15
◢ Cleaning solution (%)	Continuous	4	10
◢ Rinse time (min)	Continuous	1	3

FIGURE 9.4
High and low settings for process input factors from JMP screen shot.

		Pattern	Ultrasonic bath time (min)	Cleaning solution (%)	Rinse time (min)	Sodium ions (atoms/cm^2)
1	--+		5	4	3	•
2	---		5	4	1	•
3	-++		5	10	3	•
4	-+-		5	10	1	•
5	+-+		15	4	3	•
6	+--		15	4	1	•
7	+++		15	10	3	•
8	++-		15	10	1	•

FIGURE 9.5
List of eight experimental runs generated from JMP software.

Factors

Continuous	Discrete numeric ▼	Categorical ▼	Remove	Add n factors	1

Name	Role	Values		
◢ Ultrasonic bath time (min)	Continuous	−1		1
◢ Cleaning solution (%)	Continuous	−1		1
◢ Rinse time (min)	Continuous	−1		1

(a)

	Pattern	Ultrasonic bath time (min)	Cleaning solution (%)	Rinse time (min)	Sodium ions (atoms/cm²)
1	---	−1	−1	−1	•
2	--+	−1	−1	1	•
3	-+-	−1	1	−1	•
4	-++	−1	1	1	•
5	+--	1	−1	−1	•
6	+-+	1	−1	1	•
7	++-	1	1	−1	•
8	+++	1	1	1	•

(b)

FIGURE 9.6
Coded high and low settings for process variables: (a) for the high and low settings for our key process variables and (b) the eight experimental runs of the experiment.

the calculations. There is no need for us to deal with coded or scaled values if we are using JMP or another statistical software package.

Once we have the design, we can run the experiments and compile the results. (Okay, I realize this may take a long time to do and collect all the data, but I want to focus on the analysis here.) The results are shown in Figure 9.7. Before beginning any mathematical analysis, we want to review the data to make sure that everything seems reasonable. In this case, I observe that the longer ultrasonic bath time reduces the sodium by half. The longer rinse time also appears to help. However, the cleaning solution is not so straightforward. Since this appears to be a reasonable result and

	Pattern	Ultrasonic bath time (minutes)	Cleaning solution (%)	Rinse time (minutes)	Sodium ions (10¹² atoms/cm²)
1	---	−1	−1	−1	35
2	--+	−1	−1	1	30
3	-+-	−1	1	−1	28
4	-++	−1	1	1	26
5	+--	1	−1	−1	9.5
6	+-+	1	−1	1	7.3
7	++-	1	1	−1	15.5
8	+++	1	1	1	13.6

FIGURE 9.7
Coded runs for the full factorial with results tabulated in JMP.

nothing looks out of the ordinary, let's calculate the effects of each of the key process variables. To do this, we simply sum the sodium result values when the ultrasonic bath time is high and subtract the sum of the sodium values when the ultrasonic bath time is low, dividing the results by 4. The effect of ultrasonic bath time:

$$\frac{(9.5+7.3+15.5+13.6)-(35+30+28+26)}{4}=-18.3$$

What does this mean? When the ultrasonic bath time is set at the high level (15 minutes) the process removes 18.3 more sodium ions on the parts, as opposed to the low level (5 minutes). In other words, when we increase the ultrasonic bath time from 5 minutes to 15 minutes, we reduce the sodium ions by 18.3×10^{12} atoms/cm^2 on the parts. All of this yield improvement can be attributed to ultrasonic bath time alone since, during the four high ultrasonic bath time experiments, the other two input factors were twice low and twice high.

Now, let's look at the main effect of cleaning solution

$$\frac{(28+26+15.5+13.6)-(35+30+9.5+7.3)}{4}=0.3$$

and the effect of rinse time

$$\frac{(30+26+7.3+13.6)-(35+28+9.5+15.5)}{4}=-2.8.$$

The effect of increasing the *cleaning solution* from the low level to the higher level results in an increase in the sodium ions by 0.3×10^{12} atoms/cm^2. Increasing the *rinse time* from the low level to the higher level reduces the sodium ions by 2.8×10^{12} atoms/cm^2.

Now we want to calculate the interaction terms. We'll use our coded matrix again for this. The coded value for the interaction terms is the product of the two coded values for the main input factors for each run. To calculate the coded value for the interaction term *ultrasonic bath time * cleaning solution*, we need to multiply the coded values for each of these input factors. The results can then be displayed in a new column that represents the coded values of the interaction term. Figure 9.8 shows three additional columns,

	Pattern	Ultrasonic bath time (minutes)	Cleaning solution (%)	Rinse time (minutes)	Ultrasonic bath time * cleaning solution	Ultrasonic bath time * rinse time	Cleaning solution * rinse time	Sodium ions (10^{12} atoms/cm²)
1	---	-1	-1	-1	1	1	1	35
2	--+	-1	-1	1	1	-1	-1	30
3	-+-	-1	1	-1	-1	1	-1	28
4	-++	-1	1	1	-1	-1	1	26
5	+--	1	-1	-1	-1	-1	1	9.5
6	+-+	1	-1	1	-1	1	-1	7.3
7	++-	1	1	-1	1	-1	-1	15.5
8	+++	1	1	1	1	1	1	13.6

FIGURE 9.8

Coded runs for the full factorial with yield results and interaction factors.

one for each two factor interaction. For example, let's look at run 1. The coded value for *ultrasonic bath time* is −1 and the coded value for *cleaning solution* is −1. The product of these two gives us the value for the interaction term.

$$Ultrasonic\ Bath\ Time * Cleaning\ Solution$$

$$= (Ultrasonic\ Bath\ Time) \times (Cleaning\ Solution)$$

$$= (-1) \times (-1) = 1$$

Now we'll use the same method to calculate the effect on yield of the interaction terms. The interaction term physically means the change in sodium atoms present when the *ultrasonic bath time* and *cleaning solution* values are both low or are both high, as opposed to when one is high and the other is low. The effect of the interaction term *ultrasonic bath time * cleaning solution* can be calculated.

$$\frac{(35+30+15.5+13.6)-(28+26+9.5+7.3)}{4} = 5.8.$$

The effect of the interaction term *ultrasonic bath time * rinse time* is

$$\frac{(35+28+7.3+13.6)-(30+26+9.5+15.5)}{4} = 0.7.$$

The effect of the interaction of *cleaning solution concentration * rinse time* is

$$\frac{(35+26+9.5+13.6)-(30+28+7.3+15.5)}{4} = 0.8.$$

While I've not included this term in Figure 9.8, we can also calculate the effect of the interaction of all three *ultrasonic bath time * cleaning solution concentration * rinse time* in a similar manner to the other interaction terms:

$$\frac{(30+28+9.5+13.6)-(35+26+7.3+15.5)}{4}=-0.68.$$

In this example, most of the interactions have little effect on the sodium ions remaining after cleaning. The *ultrasonic bath time * cleaning solution concentration* interaction term shows a reduction in sodium ions by 5.8×10^{12} atoms/cm^2 when either both input factors are at their lowest level or both input factors are at their highest level. Comparison of the relative values of each main effect and interaction terms tells us that the most significant in order are *ultrasonic bath time, ultrasonic bath time * cleaning solution concentration, rinse time,* and *cleaning solution concentration * rinse time.* Notice that the interaction terms were more significant than the main effect of *cleaning solution concentration.* However, the interaction term *ultrasonic bath time * cleaning solution concentration* tells us that the *cleaning solution concentration* is an important process (input) variable in reducing the sodium ions from the surface of the parts.

9.8 FRACTIONAL FACTORIAL EXAMPLE

In some situations, experiments can be costly, either in money, time, or other precious resources. We might decide to conduct fewer experiments. Let's run through the analysis of the previous experiment assuming that we performed a fractional factorial instead of the full factorial. Instead of a 2^3, we ran a 2^{3-1}. Figure 9.9 shows the experiment generated by JMP with the results from each of the experimental runs.

Table 9.2 shows the confounding pattern for this fractional factorial example. Since the interaction terms are confounded with the main effects in a fractional factorial, we only have to calculate the main effects of the key process variables. The *ultrasonic bath time* main effect in this case is

$$\frac{(9.5+13.6)-(28+30)}{2}=-17.5.$$

	Pattern	Ultrasonic bath time (minutes)	Cleaning solution (%)	Rinse time (minutes)	Sodium ions (10^{12} atoms/ cm^2)
1	-+-	-1	1	-1	28
2	+--	1	-1	-1	9.5
3	+++	1	1	1	13.6
4	--+	-1	-1	1	30

FIGURE 9.9

Coded runs for the fractional factorial with yield results. In this case, interactions are confounded with the main effects.

TABLE 9.2

Confounding Pattern for the Fractional Factorial Example

Effects	Aliases
Ultrasonic bath time (min)	= Cleaning solution (%) * rinse time (min)
Cleaning solution (%)	= Ultrasonic bath time (min) * rinse time (min)
Rinse time (min)	= Ultrasonic bath time (min) * cleaning solution (%)

The *cleaning solution* main effect is now

$$\frac{(28+13.6)-(9.5+30)}{2}=1.1.$$

The *rinse time* main effect is

$$\frac{(13.6+30)-(28+9.5)}{2}=3.1.$$

Table 9.3 summarizes the results from both the full and fractional factorial. Notice that the main effect results are not the same as the full factorial experiment. Recall in the full factorial, we found that one of the interactions was significant. Here in the fractional factorial, we have no way of knowing if an interaction is significant. There are a number of conclusions that can be drawn with regard to the effects of these variables on the final result.

1. The ultrasonic bath time has the strongest effect in both the full and fractional factorial.
2. The cleaning solution has the smallest effect in both the full and fractional factorial.

TABLE 9.3

Calculated Effects for All Terms in the Full and Fractional Factorial Examples

Effects	Full Factorial Results	Fractional Factorial Results
Ultrasonic bath time (min)	−18.3	−17.5
Cleaning solution (%)	0.3	1.1
Rinse Time (min)	−2.8	3.1
Cleaning solution (%) * rinse time (min)	5.8	
Ultrasonic bath time (min) * rinse time (min)	0.7	
Ultrasonic bath time (min) * cleaning solution (%)	0.8	
Ultrasonic bath time (min) * cleaning solution (%) * rinse time (min)	0.8	

3. The magnitude of the effect for rinse time is similar between the full and fractional factorials but the signs are opposite. (More on this in the next section.)

What do these numbers mean? The "average" part has 20×10^{12} ions/cm^2 of sodium for the runs in this experiment. If the ultrasonic bath time is increased from our low level to the high level, we expect to reduce the sodium ions per part by 17.5×10^{12} ions/cm^2. Similarly, if the cleaning solution is increased from the low to high level, we would expect to increase the sodium ions per part by 1×10^{12} ions/cm^2. However, when the rinse time is increased from the low to high level, we would expect to increase the number of sodium ions per part by 3.1×10^{12} ions/cm^2. This is not what we see in the data or in the full factorial results. Although the strong effect of ultrasonic bath time is reasonably estimated, the other two main effects are not well captured by this fractional factorial model at all.

9.9 COMPARING FULL AND FRACTIONAL FACTORIAL RESULTS

Let's take a closer look at the full and fractional factorial models. We see in Figure 9.10 that both the full and fractional factorial models show good fit to the data.

What can we learn from this experiment? Based on these results from either of these designed experiments, we can predict the best combination of factors to improve a reduction in sodium ions. In the full factorial, the best combination of input factors, the run that gave the lowest sodium ions per part, was a long ultrasonic bath time, a long rinse time, and low cleaning solution. In the fractional factorial case, our best run in the experiment was for long ultrasonic bath time, low cleaning solution, and the shorter rinse time.

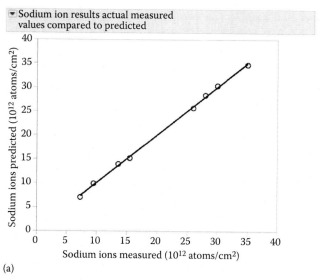

(a)

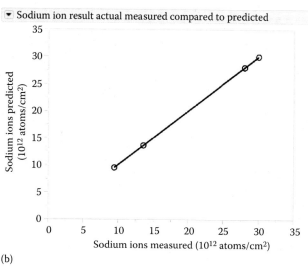

(b)

FIGURE 9.10

Comparison of actual versus predicted from the model for both the full (a) and fractional (b) factorial models.

The cube plots in Figure 9.11 provide a visual of the process space for the two sets of experiments. We can see a direct comparison between the full factorial model and predictive capability of the fractional factorial model. The fractional factorial experiment predicts that the best experimental conditions, with the lowest value for sodium ions present on the parts,

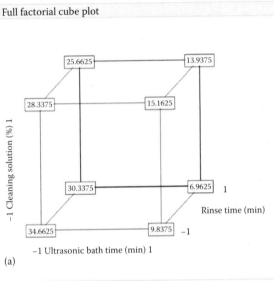

Full factorial cube plot

(a)

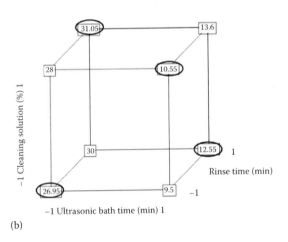

Fractional factorial cube plot

(b)

FIGURE 9.11
Cube plots of the full (a) and fractional (b) factorial process space. The "predicted" values are circled for the fractional factorial.

give us 9.5×10^{12} ions/cm^2. This was run 2 in the fractional factorial experiment. Notice that this predicted value is higher than the value we obtained as our best result in the full factorial model, where we had more experimental data on which to build our model and where our model predicted 7.0×10^{12} ions/cm^2.

Let's discuss this discrepancy between the full and fractional factorial results. The full factorial tells us that rinsing longer will reduce the sodium ions while the fractional factorial would have us use the shorter rinse time. Let's take look at the confounding pattern for *rinse time* in the fractional factorial. We see in Table 9.2 that the main effect term *rinse time* in the fractional factorial is actually confounded with the interaction term *ultrasonic bath time * cleaning solution*. From Table 9.3, recall the magnitude of this interaction term. The magnitude of the *ultrasonic bath time * cleaning solution* interaction term had the second largest effect magnitude in the full factorial. With the fractional factorial, we cannot tell the difference between the main effect and the interactions. A strong interaction effect can mask the main effect that it confounds.

9.10 NONLINEARITY, REPEATABILITY, AND FOLLOW-UP EXPERIMENTS

When we compare this to the result from the full factorial, we see that the lowest sodium ions per part were found under different conditions. Although tempting, just looking at the best run conditions may be misleading, as we can see from this example. Ultimately, with a screening experiment, we want to screen for the important factors then follow up with an experiment(s) to characterize the process space and identify an optimum condition.

It's clear that 7.3×10^{12} ions/cm^2 is better than 35×10^{12} ions/cm^2 of sodium on each part. If the specification was less than 10×10^{12} ions/cm^2 of sodium per part, for some engineers and scientists, this might be the end of the experiment. However, for others, this might be the beginning. Options for follow-up experiments might include the following:

1. Check for any nonlinearity.
2. Check for repeatability.

3. Extend the matrix even further along the *path of steepest descent.* Would additional ultrasonic bath time and rinse reduce the sodium even further?

4. Since cleaning solution had the least effect, we might consider dropping it from any further experimentation, in other words, holding it constant.

5. Since the cleaning solution had little impact on the results, we might take this as good news. Maybe we could relax controls on the cleaning solution or maybe we could further reduce the solution to 1% or 2% and save money.

There are many options for future combinations. These are only a few. I want to spend this last paragraph discussing the *path of steepest ascent/ descent.* This is a part of strategic experimentation that is outlined by Professors George Box, William Hunter, and Stuart Hunter in *Statistics for Experimenters: An Introduction to Design, Data Analysis and Model Building.* Professor Box and co-authors outline a procedure for creating contour lines using the method of least squares within the process space. The path of steepest ascent/descent is perpendicular to the contours (Box et al. 2005). Another follow-up option might be to allow the path of steepest ascent or descent to predict an extrapolated value of interest and perform several exploratory runs based on the prediction. This experimental evolution is actually a strategic use of resources that allows us to continue to improve the experimental results.

9.11 KEY TAKEAWAYS

Designed experiments allow multiple process variables to be changed simultaneously and at the same time allow us to capture complex interactions between the variables with fewer experiments. Although the ideas and concepts involved in designed experimentation are fairly straightforward, larger designs quickly become complex. Software packages make the computations and graphical preparation easy. It is critical that experimenters keep in mind that the software package doesn't analyze the meaning or interpret the results. Analysis and interpretation are still up to the investigator.

P.S. At this point, you are ready; try a simple, inexpensive designed experiment where you vary two or three process variables. Trail versions of statistical software packages can be downloaded for free, which allow easy analysis.

REFERENCES

Box, G. E. P., W. G. Hunter, and J. S. Hunter. 2005. *Statistics for Experimenters: An Introduction to Design, Data Analysis and Model Building, 2nd Edition.* New York: John Wiley & Sons.

Buie, M. J. and F. Khorasani. 1998. Using Simulation for Matrix Determination in Process Characterization. Presented at Sematech Statistical Processes Conference in Austin, TX.

Dormehl, L. 2014. *The Formula: How Algorithms Solve Our Problems...and Create More.* New York: Penguin Group.

Fisher, R. A. 1925. *Statistical Methods for Research Workers.* Edinburgh: Oliver and Boyd.

GE. 2016. What the Doctor Ordered: New Silicon Valley Startup and Stanford Health Care Will Test Digital Device Claims. GE Report. http://www.gereports.com/post /112786788335/what-the-doctor-ordered-new-silicon-valley.

Guyatt, G. 1991. Evidence-based Medicine. *ACP Journal Club* 114:A-116.

Jones, B. and C. J. Nachtsheim. 2009. Split-Plot Designs: What, Why, and How. *Journal of Quality Technology,* 41(4):340–361.

Sobel, D. 2000. *Galileo's Daughter: A Historical Memoir of Science, Faith and Love.* New York: Penguin Books.

Sur, R. L. and P. Dahm. 2011. History of Evidence-based Medicine. *Indian Journal of Urology* 27(4):487–489.

10

Strategic Design: Bringing It All Together

You've got to think about big things while you're doing small things, so that all the small things go in the right direction.

Alvin Toffler

Let's talk about planning. You may be wondering why I'm including a chapter on planning at the very back of the book rather than the very first topic. There are multiple reasons. First, planning may be the single activity that we resist the most. I've experimented and taught experimentation for several decades now. Even when I provide a template and stress how important planning is, most students and new scientists and engineers give me a "deer in the headlights" kind of stare when I want to review their plan. The second reason for including it here is because we've finally covered all the considerations for experimentation. What procedures or checklists need to be prepared? What noise factors can I live with as a part of the experiment? What do I need to measure and control? What equipment will be used? Is the equipment reliable? Is the equipment repeatable? How much systematic and random variation can I expect for each measurement? How much random variation is there within certain materials? What type of experiment would best answer my questions? What type of analysis will I need to do on the data? Everything we've covered is a piece of the puzzle; now, in this chapter, we bring it all together in a plan.

10.1   PROCESS OF PLANNING

There is a great quote from Dwight D. Eisenhower's address to the National Defense Executive Reserve Conference from November 14, 1957, about planning that I always come back to. Speaking about going to battle but the same concepts translate to experimentation, President Eisenhower says,

> I heard long ago in the Army: Plans are worthless, but planning is everything. There is a very great distinction because when you are planning for an emergency you must start with this one thing: the very definition of "emergency" is that it is unexpected, therefore it is not going to happen the way you are planning. So the first thing you do is to take all the plans off the top shelf and throw them out the window and start once more. But if you haven't been planning you can't start to work, intelligently at least. That is the reason it is so important to plan, to keep yourselves steeped in the character of the problem that you may one day be called upon to solve – or help to solve. (Eisenhower 1957)

As President Eisenhower declares, planning is an essential step in solving the problem. It is only by wrestling with the plan that we begin to get our hands around the actual problem there is to solve. With the big picture objective in mind, we can begin to fill in the details of the actions that we need to take in order to solve the problem.

I completely understand the urge to get started with solving a problem. We are eager to DO; do something, do anything. However, when we act first without planning, without preparation, we may end up with ambiguous or questionable results or misleading results. A planned and prepared experiment doesn't preclude ambiguous, questionable, or misleading results. However, if we've done all the preparation that we've discussed thus far, we have increased the odds of a successful experiment in our favor.

10.2   WHAT'S IN A PLAN?

There are lots of good reasons for developing plans prior to experimenting. The primary reason for planning is being completely prepared or as prepared as possible. Being prepared ensures that we've thoughtfully considered all the options available to us and that we are completely ready to proceed. There are many advantages to preparing and planning how to solve a problem. When a well-planned experiment/problem solving is presented, it provides transparency, a clear picture of what will transpire, the resources needed and the logic/rationale behind the experiment.

- How well the problem is understood
  - What physical/chemical/engineering/materials/etc. principles are involved
  - What have other research groups have done
  - What input other experienced scientists and engineers have
- How well the problem is articulated
  - What approaches were considered and chosen or not to pursue and why
  - How well the problem is understood (to what degree of certainty)
  - How will the solution help us
  - What the expectations are from this investment
- What the variables of interest are (Input–Process–Output diagram)
  - Which variables will be controlled
  - Which variables are noise
  - Which variables are measured
  - What assumptions are being made
- What equipment/tools/other instrumentation are needed
  - What is being measured
  - How well the inherent variation of the measurements is understood
- What data will be collected
  - How the data will be collected
  - How much data will be collected
  - How will the collected data be analyzed
  - How will the data be presented
  - How well the inherent variation in the results are understood
- What materials/chemicals/other resources will be needed to solve the problem
  - How many people and how much time will be involved
  - How much training will be required
  - How well the inherent variation in the materials/chemicals/etc. are understood
  - How long it will take to solve the problem

Having a solid plan that addresses each of these concerns will help ensure that we are addressing the problem at hand and we will know to what extent we can be certain of the results. It allows us to stay on track and focus on the big picture rather than getting lost in the details. We have documentation of all our assumptions, justifications, and expectations.

Experimental plans can be formal or informal, detailed documents or sketched outlines. At a minimum, the information to include in an experimental or problem solving plan is a problem statement, required resources, uncertainty analysis, tasks to be completed, and the schedule or timeline. A more thorough plan might additionally include cost estimates and design discussion, as well as data collection, analysis, and interpretation considerations. Independent of the type of plan, some form of experimental plan should be developed prior to taking any actions.

The less experienced the problem solver is, the more detailed the plan should be. Some companies or research groups provide problem-solving templates for scientists and engineers to use. These templates will allow us to populate the information in each section. Other companies or research groups can be less formal. As a side benefit to planning, remember, the more work we do here in the planning phase, the better prepared we will be for the final report and the less time it will take to collate, organize, and document our accomplishments.

There are formal problem-solving structures available that can be helpful in creating a plan. Check with your company about a specific format. Four of the most commonly used in industry are known as A3, 8D, DMAIC, and TRIZ. A3 problem solving was developed in Japan. A3 is the size of a standard piece of paper at Toyota in Japan. A3 problem solving outlines the problem on a single piece of paper and follows the PDCA (plan–do–check–act) outline (Matthews 2010). One advantage of using the A3 format is that it forces us to be as concise as possible because everything must fit on a single piece of paper. In 1986, a team at Ford Motor Company developed the 8D template to capture the eight disciplines of problem solving (Duffy 2014, Rambaud 2011). Drs. Arthur Jonath and Fred Khorasani have published a nice example of using the 8D problem-solving approach in the development of a medical device (Jonath and Khorasani 2011). There are many other example cases published on the Internet as well. Six Sigma's structured problem-solving methodology is known as DMAIC (Define, Measure, Analyze, Improve, and Control) and was developed at Motorola also in 1986 (Castaneda-Mendez 2012, Cudney and Agustiady 2016, Wortman et al. 2007). The TRIZ methodology was developed in Russia. The initials come from the Russian name "Teoriya Resheniya Izobreatatelskikh Zadatch," which, translated into English, is "Theory of Inventive Problem Solving" or "Creative Problem Solving Method." Genrich Altshuller, a Russian patent reviewer, came up with this method in 1946 by studying how discoveries were made (Cerit et al. 2014, Ekmekci and Koksal 2015). The TRIZ Foundation is a good resource for examples of using the method (TRIZ 2016).

We'll not go into great detail with these methodologies here (with the exception of the DMAIC technique), but they may be helpful in creating a template and structure for problem-solving planning. I have used all of these; however, because of my Six Sigma training, I naturally migrate toward the DMAIC plan. I recommend learning a bit about different problem-solving approaches to determine which might be the most valuable for a particular situation. Additionally, check with your company as they may prefer one approach over another.

Before we discuss the DMAIC planning methodology, I want to state what may be obvious to some but not so much to others. Initially, the planning document should start out very detailed in the Define and Measure sections and become less detailed in the Analyze, Improve, and Control sections. As more information becomes available during the experimentation, revamping of the planning document will be required, most likely several times. Don't be afraid to do this. Remember President Eisenhower's advice and "throw them [plans] out the window." The more information we gather and learn, the more our planning document grows. Keep the document alive and updated until the experiment is completed or the problem is solved.

10.3 DMAIC: DEFINE, MEASURE, ANALYZE, IMPROVE, CONTROL

Each letter in the DMAIC acronym represents a different phase of the problem solving. The phases are Define, Measure, Analyze, Improve, and Control. I'm not going to cover these in great depth here; references are provided in Chapter 12 for anyone who wants to read more about these phases. However, I do want to touch on the DMAIC phases briefly.

In the Define section of a planning document, it is important that we completely understand the problem and what we are trying to accomplish with our experiment. We want to make a list of everything we want out of the experiment. Are we performing this experiment or solving a problem for a manager? In either case, really nailing down the problem statement and understanding the exact nature of what's behind the problem is critical. In the problem statement, we are setting the context of the problem that we will be solving. The context is made up of the circumstances of the problem. It is worth investing time to make the content and context rich with detail. The context may be a large problem, but we may only deal with some small piece of it. For example, let's say our company is working on a new bridge design to improve earthquake resistance. Our problem may be to investigate a new fastener alloy. In the problem statement, we will want to explain the big picture but quickly narrow in to the part of the problem we will be working on. The problem statement should provide the framework which helps us focus our activities during the problem solving.

The objective of the Measure section is to identify all possible variation sources. The Measure section should provide information on the important variables. Our Input–Process–Output diagram and process flow diagram should be included in this section. It is critical to include every variable that might impact the process here. We will want to specify which equipment, instruments or tools that we will need. Also, in this section, we should outline or reference any checklists or standard operating procedures that have already been created or will be needed to insure no unwanted variation inadvertently enters our experiment. In the Measure section, we will want to identify what data need to be collected what data should be summarized. As a part of this section, we will want to perform the uncertainty analysis, i.e., determine the random and systematic variation. By performing the measurement system analysis and random variation studies as a part of the planning activities, we will be able to predict the uncertainty in our experimental results before we actually begin the experiment. Another advantage of performing the uncertainty analysis prior to beginning the experiment is that we will then be able to determine if the signal we are interested in will be present. If the bias, repeatability, reproducibility, stability, linearity, and/or random variation are so large that they dominate the results, we will be able to seek alternative equipment, instruments, tools, or materials.

In the Analyze and Improve sections of a plan, we will want to outline our plans for "intentional variation." We will specify which of the variables will vary and which will be held constant. We will use our "knowledge of the subject matter" to select process variables. With the process variables determined, it is possible to select an experimental design. Selecting the design is not as easy as it sounds, even when we know the number of variables. We want to keep the experiment as simple as possible. We never want to run an experiment that we don't completely understand. We may choose to run a one-factor-at-a-time or designed experiment or a combination of both. In these steps, we want to include as much detail as we can but until we begin the experiments, we may not know all the next steps. However, it's important to build in some time and resources for improvements to the initial experiment.

Another critical step in the experimental design selection is to confirm that all the runs are possible. When choosing the range of settings for input factors, avoid extreme values. In some cases, extreme values will give runs that are not feasible, in other cases, extreme values might move the experimental process space into some unstable region. Early in my career, I performed etch experiments frequently. The experimental variables that were critical to the etch results were often gas flow and pressure. There were settings in the design where the etch chamber might operate stably. Here, again, my knowledge of subject matter came in handy. Because I knew this was an issue, before I dedicated expensive resources to the experiment, I would confirm that my chamber, meters, and gauges operated consistently under the experimental conditions.

The final step in DMAIC problem planning is Control. In this phase of our experimentation, we are quantifying the natural variation in the solution. We want to ensure that our results are repeatable and reproducible. This step has us verifying our uncertainty analysis by putting it to the test. Just as all equipment, instruments, and tools will have some inherent variation, all processes will have inherent variation. In this section of our planning, we will build in the characterization of the inherent variation in the final process. Oftentimes, this step is omitted in design experimentation. We've spent the prior eight chapters dealing with variation (unwanted, random, systematic, and intentional) in an effort to minimize uncertainty in our results. Assuming variation has been appropriately handled, the results should be repeatable by others.

If each of these steps is completely documented, we may find creation of the final report of the experiment much simpler. The planning document easily converts to a report on the experiment, regardless of whether we are preparing an internal company memo, a journal article, or a presentation. The Define section gives us the information we need to write the Introduction section to a report. The Measure section in the planning document provides the information we need in the Materials and Methods section of a written paper. The Results, Discussion of Results, and Conclusions sections in a paper will include what we discover in the Analyze, Improve, and Control sections of the plan.

Although not required, I'd recommend having another experienced engineer review the plan prior to beginning experimentation. Review the proposed design with everyone interested in the experiment and let them review the runs. If we get buy-in ahead of time from all interested parties, everyone will be more comfortable with the decision to proceed with the plan.

Finally, in the end, when we present our results, we want to personally be confident in our data and have our intended audience trust that we've done a good job. Three good areas to focus on when sharing a plan that will boost our colleagues' confidence are data generation, known physical laws, and previous measurements or calculations of the same or a similar phenomenon. "The description of data generation is crucial. The identification and control of all relevant independent variables must be addressed and demonstrated. ... Consequently, documentation of control of the experimental conditions is the most important" (NIST 2016). Some areas of science and engineering (electronics) are well understood and the methods and measurement practices very well characterized. We expect there to be good comparison between our collected data and published data. However, in other areas of science and engineering (corrosion) where behaviors are less understood and experimental data may not compare well, we need to provide as much detail as we can in our planning and thereafter in the presentation of the data.

10.4 MURPHY'S LAW

Assuming that we've planned our experiment well, have a fully characterized measurement system, and have quantified any random variation, execution of the experiment should be straightforward. Well, don't forget about Murphy's Law. Murphy's Law roughly states that if anything can go wrong, it will ... and at exactly the wrong time. There is another form: "if someone can get it wrong, they will." No matter how much we plan, things will happen that aren't in the plan. Unexpected events occur frequently, and when they do, we will need to decide whether to proceed or start again. Equipment may break down. Hardware or processing equipment failures midexperiment may bring into question the data points just prior to failure. Samples may get contaminated. Process steps may be skipped. If we learn that process steps were omitted from the operating procedure or checklist, we must then decide to proceed or stop and evaluate. The same is true for other anomalous events; based on what we know about the results, we will need to make the call. We will definitely need to record every detail of the experiment to keep for future reference.

When things don't go according to plan or if we don't have a plan for something, there are always options. A certain amount of improvisation will happen in all experiments and problem solving. The key first step is to put whatever has happened into perspective. Ask how critical is this discovery or failure? Is there another engineer or manager who can provide assistance or advice? If so, reach out to experienced engineers or managers to determine what they might do. The next step is really to examine all the options. All that remains is to pick an option and plan with this new information in mind. Keep in mind that things are going to happen and improvisation will be necessary, but don't let it compromise all your hard work.

10.5   KEY TAKEAWAYS

Test plans are an important and sensible part of performing an experiment. Plans save time and money, assist in getting the best results, and can facilitate speedy test report writing. Reviewing a plan with a more experienced engineer prior to performing any part of the experiment may help us avoid costly mistakes. Most of all, it's important to have a good balance between planning and improvisation.

REFERENCES

Castaneda-Mendez, K. 2012. *What's Your Problem? Identifying and Solving the Five Types of Process Problems*. New York: Productivity Press/Taylor & Francis.

Cerit, B., G. Kucukyazici, and D. Sener. 2014. TRIZ: Theory of Inventive Problem Solving and Comparison of TRIZ with the Other Problem Solving Techniques. *Balkan Journal of Electrical & Computer Engineering* 2(2):66–74.

Cudney, E. A. and T. K. Agustiady. 2016. *Design for Six Sigma: A Practical Approach through Innovation*. New York: CRC Press/Taylor & Francis.

Duffy, G. L. 2014. *Modular Kaizen: Continuous and Breakthrough Improvement.* Milwaukee, WI: ASQ Quality Press.

Eisenhower, D. D. 1957. Mr. Eisenhower, the 34th President of the United States, addressed the National Defense Executive Reserve Conference on November 14, 1957. The full speech can be found online at The American Presidency Project. http://www.presidency.ucsb.edu/ws/?pid=10951.

Ekmekci, I. and M. Koksal. 2015. Triz Methodology and an Application Example for Product Development. *Procedia—Social and Behavioral Sciences* 195:2689–2698.

Jonath, A. and F. Khorasani. 2011. The 8 Disciplines Problem Solving Process: Application to a Medical Device. Paper presented at the Medical Electronics Symposium on September 27, 2011.

Matthews, D. D. 2010. *The A3 Workbook: Unlock Your Problem-Solving Mind.* New York: Productivity Press/Taylor & Francis.

NIST. 2016. National Institute of Standards and Technology website. www.srdata.nist.gov/ceramicdataportal.pds.

Rambaud, L. 2011. *8D Structured Problem Solving: A Guide to Creating High Quality 8D Reports.* Breckenridge, CO: PHRED Solutions.

TRIZ. 2016. TRIZ foundation website. www.ideationtriz.com/TRIZ_foundations.

Wortman, B., W. Richardson, G. Gee, M. Williams, T. Pearson, F. Bensley, J. Patel, J. DeSimone, and D. Carlson. 2007. *The Certified Six Sigma Black Belt Primer.* West Terre Haute, IN: The Quality Council of Indiana.

11

Where to Next?

There will be opened a gateway and a road to a large and excellent science into which minds more piercing than mine shall penetrate to recesses still deeper.

Galileo

The prior chapters have introduced the strategic problem solving concepts necessary to confidently craft experimental plans and effectively communicate findings. Chapter 2 examined the myths related to problem solving that can stop us from pushing forward. Chapter 3 reviewed the importance of communication and the common tools used in communication. The types and characteristics of data, definition of uncertainty, and an introduction to variation were presented in Chapter 4. Chapters 5 through 7 introduced three basic types of *variation* found in experimentation. Chapter 5 covered the importance of controlling *unintentional variation* with preparation of checklists and/or *standard operating procedures*. Chapter 6 explored *systematic variation* introduced by measurement equipment, while Chapter 7 looked at natural *random variation* within an experiment. *Intentional variation* was covered in Chapters 8 and 9, where the resulting data were used to build representative, descriptive mathematical models. Chapter 10 introduced the critical nature of strategic experimentation. My primary goal has been to compile the tools and organize an overarching strategy for anyone new to problem solving and experimentation and to provide additional resources and reference materials for further growth.

Where to next? All that's left to do is begin experimenting. Beginning can be the most difficult part, but once we start wrestling with these tools and strategies, doors will begin to open. It is only by struggling through these ideas and concepts and living in the uncertainty of whatever happens that experimental problem solving muscles begin to develop. It is then and only then that we begin to discover for ourselves the fascinating, amazing world of science and engineering.

A journey of a thousand miles begins with a single step.

Lao-Tzu

I've tried to introduce the ideas and concepts in an approachable way such that the material covered herein builds on the foundation of the knowledge of our science and engineering courses from our classrooms. Before closing the book, here are a few final thoughts about what it takes to be a successful experimenter and/or problem solver.

In industry today, it is difficult to get the opportunity to solve problems and experiment without the knowledge that is assumed inherent in a college degree. Our college courses expose us to the information that will presumably open doors for our future. We are taught what others have discovered and learned over the years. The recipes that we follow in our lab courses expose us to one way of solving a problem or running an experiment. In Chapter 1, we saw that knowledge of our subject matter is critical; without this foundation, we are on shaking ground. I want to reiterate how important this foundation is. You may be wondering at this point, if the knowledge of subject is so important, then why have we not discussed it at great length in this book? The answer is simple: almost all engineers and scientists graduate college with a decent handle on subject matter. Knowledge of subject matter may be stronger for some than others, but in college, we are at least introduced to the material required to become an engineer in our field. However, very few science and engineering graduates, at any level, have even a basic knowledge of strategy for problem solving.

Recall from Chapter 1 that there are two requirements for playing the game of experimentation well: knowledge of subject matter and knowledge of strategy. Knowledge of subject matter is all the material we learned in school and all the material we know from our experience. Knowledge of subject matter refers to the information we need to have ready access to, the information running in the background as we solve problems. The latest discoveries and recent developments in our field also contribute to this knowledge of subject matter. Knowledge of subject matter is both a necessary and sufficient condition for problem solving. There are many successful scientists and engineers who have not been exposed to strategic problem solving. Knowledge of strategy takes that necessary knowledge of subject matter and makes it elegant.

Imagination is more important than knowledge.

Albert Einstein

Although knowledge of our subject area and basic skills in our field are critical, in order to be a successful problem solver, we must hold on to our imagination and creativity. For scientists and engineers, knowledge and creativity go hand in hand. Knowledge of subject matter is the substance that feeds and fuels our curiosity and creativity. As young children, we might have been constantly creating. Many of us have become less creative as we get older. It is possible to recapture that curiosity and imaginative exploration by allowing ourselves to be inquisitive and curious. Experiments test out our theories and answer niggling questions. Experiments help us solve problems. To create is to be human, and as with other abilities, it can be developed into a spectrum of competencies.

Curiosity and creativity alone are not enough, however. It is through the deliberate accumulation of subject knowledge that "AHA!" moments arise. We need background information in our long-term memory in order to more be more creatively efficient. As educators have long known, breakthroughs, those "EUREKA!" moments, are more likely to spring from a larger store of background materials in our long-term memory library (Leslie 2014). To rephrase Louis Pasteur's famous statement, luck or serendipity favors the prepared, sagacious mind.

Great scientists and engineers didn't win an experiment lottery. Their most impressive discoveries did not fall into their laps. Think of the brothers Orville and Wilbur Wright (McCullough 2015). They weren't discouraged by the myriad of problems they encountered along the way to the first flight. Each problem was another hurdle that got them closer to the big problem they wanted to solve. They worked for years, dedicating their lives and sacrificing much for the accomplishment. The problem solving underlying creativity is cognitive thought with "very high degrees of persistence and motivation" (Weisberg 2013). Orville and Wilbur's persistence and motivation likely resulted from their *mindset.*

Our mindset has to do with our own self-perception about how we learn or if we can learn something new (Dweck 2007). With a fixed mindset, we cannot grow and develop further. With a growth mindset, we can.

It's not our talents or abilities or IQ or creative potential that allows us to be successful. It is our mindset. A growth mindset holds that hard work, perseverance, will bring further growth and development. No, it's not always easy, but a positive attitude and a stick-to-it attitude are the first steps toward achieving real creative progress. Perseverance, or grit, is more important than intelligence (Duckworth 2016, Dweck 2007, Oakley 2014).

Perseverance is a habit that we can develop. It is or can be self-taught. Madame Marie Curie's biographer, historian and author, Barbara Goldsmith, wrote of the young Manya Sklodowski (the future Marie Curie), "At eighteen, she had already 'acquired the habit of independent work': to draw her own conclusions, without the restraints of accepted perceptions. ..." (Goldsmith 2005). The earlier we make our mind up to work hard, the earlier we can begin solving interesting problems.

Choosing to persevere doesn't mean that we'll always get it right. My favorite yoga instructor says at least once in every class, "Remember, yoga is a practice. If you fall out of a pose, smile and try again." The smile really works. It is difficult to beat myself up while I'm smiling. I no longer beat myself up on those days when I can't hold the dancer or tree pose. What if we brought this attitude to our experiments? What if we looked at all experiences in life and in the lab as opportunities for growth? With each failed experiment, ask "What did I learn?" and "How can I use it to grow?"

> One must learn by doing the thing; though you think you know it, you have no certainty until you try.
>
> **Sophocles**

The first or second or third time we do anything, it is probably not going to be great. The first song on the violin will make our parents proud but probably will not get us a standing ovation at the Kennedy Center. Our first oil on canvas might make it into our parent's living room but probably not into the San Francisco Museum of Modern Art. Why would we expect this with our experiments?

Problem solving and experimentation in the sciences and engineering are creative processes. Dr. Kevin Ashton, pioneer in radiofrequency identification, compares the creative process to sketching,

> The main virtue of a first sketch is that it breaks the blank page. It is the spark of life in the swamp, beautiful if only because it is a beginning. ... When we envy the perfect creations of others, what we do not see, what

we by definition cannot see, and what we may also forget when we look at successful creations of our own, is everything that got thrown away, that failed, that didn't make the cut. When we look at a perfect page, we should put it not on a pedestal but on a pile of imperfect pages, all balled or torn, some of them truly awful, created only to be thrown away. This trash is not failure but foundation, and the perfect page is its progeny. (Ashton 2015)

Professor Henry Petroski, Duke University professor of engineering, compares engineering to writing:

Some writers even if they do not try to publish them, do not crumple up false starts or their failed drafts. They save every scrap of paper as if they recognize that they will never reach perfection and will eventually have to choose the least imperfect from among all their tries. These documents of the creative process are invaluable when they represent the successive drafts of a successful book or any work of a successful writer. ... Creating a book can be seen as a succession of choices and real or imagined improvements. (Petroski 1982)

The seesaw process of creation, of fail, revise, fail, revise, on and on, is common to art, writing, and experimentation. Each failure is an opportunity to learn and revise for the next round of experimentation.

Experience is a brutal teacher but my god how you learn.

C. S. Lewis

Great scientists have great failures. Newton, one of the greatest physicists in recorded history, spent a third of his life working on alchemy. Alfred Russell Wallace, a contemporary of Charles Darwin and independent developer of the theory of evolution, participated in experiments to communicate with the dead. Urbain Jean Joseph Le Verrier, the discoverer of Neptune, also predicted the existence of another planet, which was incorrect. Albert Einstein couldn't reconcile the fundamental controlling laws of quantum mechanics with his own intuitive understanding of his beliefs. Thomas Edison has some 1093 US patents. He also has 500 to 600 that failed. Here are a few quotes from Edison about failure:

- "I have not failed. I've just found 10,000 ways that don't work."
- "I'm not discouraged, because every wrong attempt discarded is another step forward."

- "Results! Why man, I have gotten lots of results. I know several thousand things that won't work."
- "When I have fully decided that a result is worth getting I go ahead of it and make trial after trial until it comes."
- "Many of life's failures are men who did not realize how close they were to success when they gave up."

The chemical lubricant WD-40 got the 40 in its name because the prior 39 formulations didn't work. Syphilis cure Salvarsan 606 was aptly named for the number of attempts to get it right. Linus Pauling, one of the few scientists to receive two Nobel Prizes, said, "The best way to have a good idea is to have lots of ideas." Dr. Pauling knew that it takes many, many dead ends to find a path that works.

From great scientists and engineers to newbies, we all stand on the shoulders of many who came before us. Galileo, Isaac Newton, Marie Curie, Linus Pauling, and Albert Einstein all devoted their lives to solving a big problem. They were aided by the efforts of others. The small problems had to be solved first before the big one could be addressed. For example, before Einstein could come up with theory of relativity, Faraday needed to firmly establish the relationship between electricity and magnetism, which provided the basics for the electric engine and the concept of energy. Newton gave us the property of matter known as mass. Lavoisier, the father of chemistry, showed us how the mass of materials could be combined and separated establishing conservation of mass. Galileo provided an early experimental attempt to measure the speed of light. It was finally James Clerk Maxwell who helped us understand the relationship between electricity, magnetism, and light following the work of Danish astronomer Ole Roemer and Faraday. Robert Recorde, an English textbook publisher, gave us the equal sign. Newton and Leibniz both developed calculus in order to explain physical phenomena they observed. Newton gave us mass times velocity and Leibniz gave us mass times velocity squared. It was Emilie du Châtelet who finally settled the issue and gave us the square. Einstein harnessed the contributions of all these scientists to develop his famous equation, $E = mc^2$ (Bodanis 2000).

You might say that all that we will learn in our early problem solving or experimentation is already known by others, and you might be correct. We shouldn't let this discourage us or cause us to say, "Oh, there's nothing new here. It's no big deal. I'm just a novice engineer and what I do doesn't really matter." We can't sell ourselves short. Scientists and engineers readily admit they cannot get everything right and often mistakes or wrong information gets published. Too often, we take the word of well-respected journals or academicians and never duplicate their results. It is important that experiments be repeated—not just to verify the results of others but more importantly so that we can discover the results for ourselves. Replication of experiments is a wonderful way for us to discover for ourselves what these others before us have discovered. These experiments allow us to learn what's important and determine where improvements or permutations of the experiment might be of interest. When we discover for ourselves, this discovery is then ours.

As we gain confidence and delve deeper into the mysteries of science and engineering, there is less and less certainty. Any uncertainty created at the beginning of the movie is resolved within 120 minutes. In science and engineering, many of the mysteries we encounter may never be resolved. The only certainty is uncertainty. In school, science and engineering problems are presented as nice little packages of answers. Dr. Freeman Dyson, retired Princeton theoretical physicist, expert in quantum electrodynamics and author, observed that science is not a collection of truths but a "continuing exploration of mysteries" (Leslie 2014). What we know from science and engineering today are actually answers to the scientific puzzles, the mysteries, solved by those who came before us. Solving small pieces of the larger scientific puzzle that is life can be rewarding and at the same time provoke many more questions. For many of us, this is part of the fun of science—this never-ending quest to put pieces of the larger puzzle together—to find those puzzle pieces that aren't Google-able. This "rigorous and persistent exploration of what we don't know" is really what keeps us curious. The inventor and audio pioneer Ray Dolby said, "To be an inventor, you have to be willing to live with a sense of uncertainty, to work in the darkness and grope toward an answer, to put up with the anxiety about whether there is an answer" (Leslie 2014). As scientists and engineers, we are explorers, adventurers, and innovators each time we discover something unknown to ourselves. The more we embrace the unknown and solve problems, the easier it gets. "Part of being able to tackle complex and difficult questions is accepting that there is nothing wrong with not knowing. People who are good at questioning are comfortable with uncertainty" (Berger 2014). Each time we discover for ourselves, we become more confident through our experiences with experimentation. Questioning and experimenting go hand in hand.

Problem solving is an essential component of science and engineering. As scientists and engineers, we spend our careers solving one problem after another. The more strategic we are with our work, the more opportunities will be available to us. Our science and engineering classes give us the knowledge and background information. This knowledge can function as the building blocks in our career. However, approaching each problem strategically allows us to operate more efficiently. By rediscovering our curiosity and imagination, we can reconnect with our creative longing to discover for ourselves. Perseverance and passion will be required in the face of null results, uncertainty, and what looks like failure. We begin by imitating other experiments (if we can) or by verifying known information so that we discover for ourselves what others have already learned. As we gain facility and become more knowledgeable about experimentation and problem solving, we can venture into new territory.

May you enjoy the process, fearlessly face the uncertainty with wonder and curiosity, and discover for yourself. Happy experimenting!

REFERENCES

Ashton, K. 2015. *How to Fly a Horse: The Secret History of Creation, Invention, and Discovery*. New York: Doubleday.

Berger, W. 2014. *A More Beautiful Question: The Power of Inquiry to Spark Breakthrough Ideas*. New York: Bloomsbury USA.

Bodanis, D. 2000. $E = mc^2$: *A Biography of the World's Most Famous Equation*. New York: Walker & Company.

Duckworth, A. 2016. *Grit: The Power of Passion and Perseverance*. New York: Simon & Schuster.

Dweck, C. S. 2007. *Mindset: The New Psychology of Success*. New York: Random House.

Goldsmith, B. 2005. *Obsessive Genius: The Inner World of Marie Curie*. New York: W. W. Norton.

Leslie, I. 2014. *Curious: The Desire to Know and Why Your Future Depends on It*. New York: Basic Books.

McCullough, D. 2015. *The Wright Brothers*. New York: Simon & Schuster.

Oakley, B. 2014. *A Mind for Numbers: How to Excel at Math and Science (Even if You Flunked Algebra)*. New York: Jeremy P. Tarcher/Penguin.

Petroski, H. 1982. *To Engineer Is Human: The Role of Failure in Successful Design*. New York: Vintage Books/Random House.

Weisberg, R. W. 1993. *Creativity: Beyond the Myth of Genius*. New York: W. H. Freeman & Co.

12

One More Thing...

... like any skill, becoming very good at scientific reasoning requires both practice and talent. But becoming tolerably good requires mainly practice and only a little talent. And for most people tolerably good is good enough. So work at developing your skills little by little.

Ronald N. Giere

The ideas and concepts gathered in this book are from my own experiences. However, in my research and preparation for this book, I discovered so many valuable and wise references from a variety of fields. Each new reference I found sent me in multiple directions to read additional new authors. I am grateful to them all for the valuable contribution they have made to the body of literature available on the topics discussed here but also to my work. The following list contains the books and papers I enjoyed the most.

12.1 REFERENCES ON EXPERIMENTATION

A wonderful reference with fun experiments that makes great reading is *Thinking, Fast and Slow* by Daniel Kahneman. Professor Kahneman is a behavioral economist. His book is an easily read collection of his experiments. Even for those of us in the physical sciences, reading about experiments in other areas can be a source of inspiration and enjoyment. Behavioral economists have fun with their experiments and it shows. Several other examples in my library referenced herein include *Predictably Irrational* by Dan Ariely and *Freakanomics* by Steven D. Levitt and Stephen J. Dubner.

12.2 REFERENCES ON COMMUNICATION

Professor Roald Hoffman's, Nobel Laureate in chemistry, writings tell me that he loves science. Jeffrey Kovac and Michael Weisberg collected his writings in, *Roald Hoffman on the Philosophy, Art, and Science of Chemistry.* I can't write this without giving physics equal time. Professor Richard Feynmann's books are also enjoyable. *The Pleasure of Finding Things Out* is a great first read for someone new to Feynmann. For additional information on the topic of data displays, Stephen Few and Professor Edward Tufte's books are invaluable. They are all wonderful and stress the codependence of language and graphics in communication. The other book that I'd recommend is Carmine Gallo's *Talk Like TED: The 9 Public-Speaking Secrets of the World's Top Minds.* Public speaking takes practice, if you do not have opportunities through your activities to present to an audience; Toastmasters International is a wonderful resource. There are many resources available to assist people with presentation skill development. TED talks are great resources available on the Internet to see great speakers—some well known and others less well known. Additional resources are available at Toastmasters, an educational nonprofit specifically established to assist members with public speaking and communication. Regional clubs have been established all around the world.

Jeffrey and Laurie Ford have written a powerful book on communication. *The Four Conversations: Daily Communication That Gets Results* is a useful tool for further development of ourselves in requesting or questioning conversations with our assistants, peers, or managers.

12.3 REFERENCES ON ERROR ANALYSIS

One of the other key concepts in this chapter that deserves more attention as engineers and scientist mature in their experimental sophistication is uncertainty characterization. The topic was touched on here. However, there are two outstanding references that should be used to understand this area further: Dr. David Deardorff from the Department of Physics at University of North Carolina Chapel Hill, with multiple references on the

Internet, and University of Colorado Physics Professor John Taylor, *An Introduction to Error Analysis*. John Taylor's book is excellent but does not necessarily conform to the methodology of the Guide to the Expression of Uncertainty in Measurement (GUM). Therefore, be careful if you are going back and forth between Deardorff and Taylor. The terms used are not always the same. Deardorff presents a readable text on uncertainty that is consistent with both National Institute of Standards and Technology (NIST) and GUM standards. The NIST and GUM standards are available on the Internet for free but much less readable. Should you find this kind of thing riveting, the related documents International Organization for Standardization GUM, NIST GUM, and the corresponding American National Standard ANSI/NCSL Z540-2 may keep you up at night.

12.4 REFERENCES ON CHECKLISTS

I found Dr. Atul Gwande's *Checklist Manifesto: How to Get Things Right* an invaluable reference book on this topic. A fascinating story that covers the development of good lab practices can be found in Rebecca Skloot's book *The Immortal Life of Henrietta Lacks*. Time-dependent measurements cannot always be avoided, and when they can't, I'd recommend beginning with either *Experimental Methods for Engineers* by J. P. Holman or *Experimentation and Uncertainty Analysis for Engineers* by Hugh W. Coleman and W. Glenn Steele.

12.5 REFERENCES ON MEASUREMENTS

As far as I know, nothing completely entertaining has been written about measurement system analysis. Therefore, I will just refer you to the measurement system analysis manual, and if that isn't enough, try reading through *Design and Analysis of Gauge R&R Studies: Making Decisions with Confidence Intervals in Random and Mixed ANOVA Models (ASA-SIAM Series on Statistics and Applied Probability)* by Richard K. Burdick, Connie M. Borror, and Douglas C. Montgomery.

12.6 REFERENCES ON RANDOMNESS

Randomness is fascinating, and like chaos, there have been many wonderful books written on the topic. A few of the books that I've really enjoyed include *Naked Statistics: Stripping the Dread from the Data* by Charles Wheelan and *The Drunkard's Walk: How Randomness Rules Our Lives* by Leonard Mlodinow. Although these books are written by super smart professors from Dartmouth and California Institute of Technology, they've been able to write about random statistical phenomena with an entertaining and historical slant. I found *Creativity, Inc.* by Ed Catmul enjoyable because he wrote on the randomness of how creative work progresses. If you are looking for a relaxing book filled with advice from a cartoonist, then Scott Adams' *How to Fail at Almost Everything and Still Win Big: Kind of the Story of My Life* is worth a read. Mr. Adams' book encourages systems thinking and looking for patterns in life. To gain a more technical understanding of the intricacies of normal distribution, one of the best resources that I've found is *Introduction to Error Analysis: The Study of Uncertainties in Physical Measurements* by John R. Taylor. Professor Taylor covers the normal distribution and develops proofs of what we know about normal distributions. He also has a chapter dedicated to Chauvenet's criterion. His book would be a great asset to an experimental physical scientist's library.

12.7 REFERENCES ON STATISTICS AND DESIGNED EXPERIMENTATION

Naked Statistics: Stripping the Dread from the Data by Professor Charles Wheelan is an excellent introductory reference on regression analysis. His examples can be ridiculous and may leave you groaning, but you will smile and maybe even laugh on occasion. How many other statistics books can you say that about? Other good references for one-factor-at-a-time experimentation are the early algebra and calculus books. The bible on designed experimentation is *Statistics for Experimenters: An Introduction to Design, Data Analysis, and Model Building* by George E. P. Box, William G. Hunter, and J. Stuart Hunter. You may find the book difficult to navigate, but stick with it. Another text that might be more approachable is

Design and Analysis of Experiments by Douglas C. Montgomery. This book is currently on its eighth edition and has a student solutions manual if you want to work through the problems. The introductory book *DOE Simplified, Practical Tools for Effective Experimentation, 3rd Edition,* by M. J. Anderson and P. J. Whitcomb can give you an easy-to-follow introduction. A good reference book is available from NIST entitled *NIST/ SEMATECH e-Handbook of Statistical Methods,* which is available online at http://itl.nist.gov/div898/handbook/index.htm. There are a number of good books on learning to use JMP available on their website: www .jmp.com.

12.8 REFERENCES ON CURIOSITY, CREATIVITY, AND FAILURE

The books *How to Fly a Horse* by Kevin Ashton, *Curious: The Desire to Know and Why Your Future Depends on It* by Ian Leslie, and *A More Beautiful Question: The Power of Inquiry to Spark Breakthrough Ideas* by Warren Berger solidified my thoughts on creativity and genius. Professor Carol Dweck's *Mindset* and Professor Barbara Oakley's *A Mind for Numbers* confirmed my own experience with research and data that perseverance is critical to success. Scott Berkun's *The Myths of Innovation* does a great job of expelling many of the myths surrounding the creative process. Professor Henry Petroski's articles and books are educational and enjoyable.

Books about great scientists and engineers who failed are fascinating. It reminds me how difficult it is and what it takes to achieve great things. *The Wright Brothers* by David McCullough is an incredible read. *Brilliant Blunders* by Mario Livio and *Einstein's Mistakes: The Human Failings of Genius* by Hans C. Ohanian are good places to begin.

In Gratitude

I gratefully acknowledge the contributions of the following friends and colleagues:

My mentor, friend, teacher, guide, hiking partner: Fred Khorasani, an incredible statistician and an amazing human being. It was through his guidance and hours (and hours and hours) of patient conversation that these statistical ideas and strategies began to take hold long ago. To Karen, his wife, for having the patience of a saint, and feeding us as we discussed this for hours at the kitchen table.

My professors, science mentors, teachers: Mary Brake, Eugene Omasta, BJ Bateman, Ron Gilgenbach, Ward Getty, Fred Terry, James Holloway, and Mitch Pindzola.

My amazing friends who read and reviewed chapters and discussed material with me: Emily Allen, Caryl Athanisiu, Jennie Brook, Stefanie Harvey, Cindy Brooks, Mary Walker, Chris Fields, Laurel Perry, Cat Ley, Amy Bunker, Rich Schuster, Michelle and George Paganini, and Stacy Gleixner.

My former students who have lived through the evolution of this work. There are far too many to mention them all by name.

The companies that I've been privileged to work for over the years who've provided the experiences that have helped solidify these ideas and concepts. In particular, the people who have made it all possible: Paul Ginouves, director of marketing communications at Coherent, Inc.; John Ambroseo, chief executive officer of Coherent, Inc.; and Mike Welch, retired, vice president of Applied Materials.

To a company that I've not had the privilege to work for but have admired and respected for years: SAS Institute, makers of JMP. In particular, Anne Milley, Laura Higgens, Curt Hinrichs, and John Sall.

A very special group of teachers and friends: Helen Gilhooley, Werner Erhardt, Jeri Echiverra, Michael Jenson, Wiley (Chip) Souba, Nancy Carney, and Marci Feldman have also made this possible in other ways. Werner spoke directly to me when he said, "Discover for yourself" in a room of 200 other people and this work began.

This work would not be where it is today without the contributions of a very patient and talented artist, Natalya Shishkina. To a very dear man, David Axelrod, for reading through the manuscript multiple times. I am grateful for his feedback and encouragement.

Finally, many thanks to Michael Sinocchi, Alexandria Gryder and the many others who made this possible at Taylor & Francis/CRC Press/ Productivity Press.